AF581709

# Hide and Seek of Soil Microbes—Who Is Where with Whom and Why?

*Editors*
Maraike Probst
University of Innsbruck
Austria

Judith Ascher-Jenull
University of Innsbruck
Austria

*Editorial Office*
MDPI
St. Alban-Anlage 66
4052 Basel, Switzerland

This is a reprint of articles from the Special Issue published online in the open access journal *Applied Sciences* (ISSN 2076-3417) (available at: https://www.mdpi.com/journal/applsci/special_issues/soil_microbes).

For citation purposes, cite each article independently as indicated on the article page online and as indicated below:

| LastName, A.A.; LastName, B.B.; LastName, C.C. Article Title. *Journal Name* **Year**, *Volume Number*, Page Range. |
|---|

**ISBN 978-3-0365-5149-4 (Hbk)**
**ISBN 978-3-0365-5150-0 (PDF)**

Cover image courtesy of Judith Ascher-Jenull

# Hide and Seek of Soil Microbes—Who Is Where with Whom and Why?

Editors

**Maraike Probst**
**Judith Ascher-Jenull**

MDPI • Basel • Beijing • Wuhan • Barcelona • Belgrade • Manchester • Tokyo • Cluj • Tianjin

# Contents

About the Editors . . . . . . . . . . . . . . . . . . . . . . . . . . . . . . . . . . . . . . . . . . . . . . . . vii

**Maraike Probst and Judith Ascher-Jenull**
Special Issue on 'Hide and Seek of Soil Microbes—Who Is Where with Whom and Why?'
Reprinted from: *Appl. Sci.* **2022**, *12*, 7693, doi:10.3390/app12157693 . . . . . . . . . . . . . . . . . . 1

**Maria Zottele, Johanna Mayerhofer, Hannah Embleton, Katharina Wechselberger, Jürg Enkerli and Hermann Strasser**
Biological *Diabrotica* Management and Monitoring of *Metarhizium* Diversity in Austrian Maize Fields Following Mass Application of the Entomopathogen *Metarhizium brunneum*
Reprinted from: *Appl. Sci.* **2021**, *11*, 9445, doi:10.3390/app11209445 . . . . . . . . . . . . . . . . . . 5

**Stavros D Veresoglou, Leonie Grünfeld and Magkdi Mola**
Micro-Landscape Dependent Changes in Arbuscular Mycorrhizal Fungal Community Structure
Reprinted from: *Appl. Sci.* **2021**, *11*, 5297, doi:10.3390/app11115297 . . . . . . . . . . . . . . . . . . 19

**Chakriya Sansupa, Sara Fareed Mohamed Wahdan, Shakhawat Hossen, Terd Disayathanoowat, Tesfaye Wubet and Witoon Purahong**
Can We Use Functional Annotation of Prokaryotic Taxa (FAPROTAX) to Assign the Ecological Functions of Soil Bacteria?
Reprinted from: *Appl. Sci.* **2021**, *11*, 688, doi:10.3390/app11209445 . . . . . . . . . . . . . . . . . . . 31

**Edoardo Mandolini, Maraike Probst and Ursula Peintner**
Methods for Studying Bacterial–Fungal Interactions in the Microenvironments of Soil
Reprinted from: *Appl. Sci.* **2021**, *11*, 9182, doi:10.3390/app11199182 . . . . . . . . . . . . . . . . . . 49

**Nieves Barros**
Thermodynamics of Soil Microbial Metabolism: Applications and Functions
Reprinted from: *Appl. Sci.* **2021**, *11*, 4962, doi:0.3390/app11114962 . . . . . . . . . . . . . . . . . . . 71

# About the Editors

**Maraike Probst**

Maraike Probst is a postdoctoral researcher at the Institute of Microbiology of the University of Innsbruck, Austria. Maraike obtained her first degree from the University of Kiel, Germany. She is a microbiologist and molecular cell biologist. Her Ph.D., obtained from the University of Innsbruck, focused on applied microbiology and molecular microbial ecology. In her Ph.D. thesis, she applied natural microbial communities for the production of lactic acid from renewable resources. The final goal of laying the foundation for a naturally friendly and ecologically efficient way of producing poly-lactic acid was achieved. Since her postdoctoral research, Maraike focuses more on basic research. During her postdoctoral period at the Ben-Gurion University, Be'er Sheba, Israel, she studied microbial ecogenomics, focusing on horizontal gene transfer in complex, natural microbial communities. The main aim was to shed some light on the hosts, movement and mobility of plasmids in natural communities. As the plasmid distribution and genetic content are considered highly relevant for the genetic diversity of microbes and microbial populations, this research asked fundamental questions regarding life, ecology and evolution.

Currently, Maraike is researching microbiomes of different ecosystems, focusing on microbial interactions and their relation to a (changing) environment.

*Research-Topics*: molecular microbial ecology; microbial communities; microbial interactions; microbial evolution and adaptation; modeling of ecological data

**Judith Ascher-Jenull**

Judith Ascher-Jenull is a soil microbiologist, postdoctoral researcher, and senior scientist at the University of Innsbruck's (A) Working Group Microbial Resource Management. Her expertise is in molecular environmental microbiology with a focus on soil and deadwood dynamics within the context of climate change. She earned her master's in microbiology from the Department of Microbiology at the University of Innsbruck, and her Ph.D. in agricultural chemistry from the University of Milan (I), She worked for 15 years as a post-doc researcher in the Dept. of Soil Science and Plant Nutrition at the University of Florence (I). She is currently also involved in the Science Centre project MikrobAlpina–Mikromondo, and since 2014, is active as Editor-in-Chief of the journal *Applied Soil Ecology*.

*Research-Topics*: molecular microbial ecology; soil microbial communities; fate and ecological relevance of soil DNA: extracellular DNA vs. intracellular DNA vs. total DNA (environmental DNA); optimization of environmental DNA extraction/quantification protocols

MDPI

*Editorial*

# Special Issue on 'Hide and Seek of Soil Microbes—Who Is Where with Whom and Why?'

**Maraike Probst * and Judith Ascher-Jenull ***

Department of Microbiology, University of Innsbruck, Technikerstrasse 25d, 6020 Innsbruck, Austria
* Correspondence: probst.maraike@gmail.com (M.P.); judith.ascherjenull@gmail.com (J.A.-J.)

## 1. Introduction

Our question posed for and used as title of the special issue *'Hide and Seek of Soil Microbes'–Who is Where with Whom and Why?* can be considered as the central paradigm of soil microbial ecology, covering and embracing all relevant aspects and topics.

- Soil is life!
- Soil, the solid matrix of all terrestrial ecosystems: the most complex, diverse and heterogeneous ecosystem, harboring plenty of micro-niches and hot spots for microorganisms (***Who is Where?***)
- Soil, the source of life, which is subjected to environmental stressors, especially within the context of anthropogenic-driven challenges. Soil microbes, the microbial inhabitants of soil, live in complex, diverse consortia. They drive all biogeochemical processes, and they are involved in all nutrient cycles (***Who is with Whom and Why?***)

Although technical innovations boost soil scientists' opportunities to assess soil ecosystems and ask big questions, fundamental focus points remain and require attention more than ever:

- **Experimental design and sampling strategy.** Although addressed previously, there are things to remember and far more things to uncover and to take home. Spatial heterogeneity in particular is still rarely considered in the set-up of scientific experiments.
- **Microbial diversity.** Ecological hypotheses involve the observation of species number. The current methodology does not necessarily provide this information. Likewise, the controversially discussed analysis of relative abundance data needs validation and scientific awareness.
- **Functions.** Omics benefit soil science. However, understanding the soil ecosystem from a mechanistic point of view differs from what recent research has termed ´function´. Functioning implies malfunctioning, thereby reducing the soil ecosystem to a system providing a service. For comprehension, coherent and precise definitions are indispensable. In addition, the precious information provided by sophisticated methods warrants critical discussion to draw relevant conclusions.

## 2. 'Hide and Seek of Soil Microbes—Who Is Where with Whom and Why?'

In light of the above, this special issue was introduced with the aim of opening a vivid discussion on the following topics:

- Adequate experimental design for representative study of soil ecosystems;
- Possibilities and limitations of observing microbial diversity;
- Possibilities and limitations of data analysis and their impact on ecological conclusions;
- Microbial spatial heterogeneity across different scales;
- How to observe microbial niche partitioning and occupation;
- Soil ecosystems, soil microbial communities, and their ´function´. In what sense can microbes, microbial communities and ecosystems have a function?

**Citation:** Probst, M.; Ascher-Jenull, J. Special Issue on 'Hide and Seek of Soil Microbes—Who Is Where with Whom and Why?' *Appl. Sci.* **2022**, *12*, 7693. https://doi.org/10.3390/app12157693

Received: 27 July 2022
Accepted: 29 July 2022
Published: 30 July 2022

This special issue represents a concise and strong scientific reaction to our questions posed, with three Regular Full Length Research Articles ([1] cited by 1, viewed by 474; [2] cited by 1, viewed by 878, and [3] cited by 28, viewed by 2772), and two Review Articles ([4] viewed by 911; and [5] cited by 3, viewed by 1038). Various topics related to our posed questions—in general or in particular—have been addressed.

The first paper, authored by M. Zottele, J. Mayerhofer, H. Embleton, K. Wechselberger, J. Enkerli, and H. Strasser [1], presents an impressive and promising example of biological pest control of Diabrotica populations (corn rootworm, *Diabrotica v. virgifera*-Coleoptera, Chrysomelidae) using inundative mass application of *Metarhizium brunneum* BIPESCO 5 (Hypocreales, Clavicipitaceae). This case study points out the importance of a strong experimental design (long-term field studies with different cultivation techniques and infestation rates) and concludes, supported by the obtained data, that crop rotation remains the option of choice for rapid pest population decline at high pest densities [1].

In the second paper, the authors S.D. Veresoglou, L. Grünfeld, and M. Mola, [2] address how soil spatial structures in plant mesocosms, i.e., habitat connectance and habitat quality, alter the predictability/stochasticity of the community composition of arbuscular mycorrhizal fungi (AMF). This is one of very few studies experimentally quantifying how micro-landscape structures increase the stochasticity of AMF communities. Thus, micro-landscape structures could support the persistence of less competitive species in the ecosystem. This is particularly meaningful in the case of AMF, which are poorly researched in this regard, as their presence/absence might determine the establishment of plant individuals in the ecosystem. Overall, the authors provided evidence that the community structure of AMF is less responsive to spatiotemporal manipulations than root colonization rates, which is a facet of the symbiosis that is currently poorly understood.

The third paper, authored by C. Sansupa, S.F.M. Wahdan, S. Hossen, T. Disayathanoowat, T. Wubet, and W. Purahong [3], presents, for the first time, the successful application of the FAPROTAX database, originally developed for marine ecosystems, to the soil ecosystem, providing evidence about its potential as a powerful tool for predicting ecological relevant functions of soil bacterial and archaeal taxa derived from 16S rRNA amplicon sequencing. The authors conclude that although FAPROTAX cannot predict the function of all detected taxa, it is capable of a fast-functional screening or grouping of 16S bacterial data derived from any ecosystem. It was suggested that additional datasets of both the taxonomy and functional references could further improve the database, thereby increasing the number of functionally assigned OTUs derived from 16S rRNA. The innovative aspect and scientific relevance of this study [3] is confirmed by the huge number of records (28 citations; 2772 views) within the first months after the release of our special issue.

In this special issue, next to the beforementioned case studies, there are two Review Articles, one critically discussing 'Methods for Studying Bacterial–Fungal Interactions in the Microenvironments of Soil' [4], and the other focusing on 'Thermodynamics of Soil Microbial Metabolism: Applications and Functions' [5].

The review by E. Mandolini, M. Probst and U. Peintner [4] perfectly matches with one of the Guest Editors' desired expectations, being among the driving forces for editing this special issue. The review focused on microscale variations in soil properties constraining the distribution of fungi and bacteria, and to the extent of their interactions and consequent behavior and ecological roles. The review points out that a realistic assessment and understanding of bacterial–fungal interactions is only possible by considering the spatiotemporal complexity of their microenvironments. The authors succeeded in further raising awareness of this important aspect by critically and extensively discussing possible methodologies, embracing culture-dependent and culture-independent tools along with suggesting new applications of current technologies to answer newly formulated research questions, in order to better glimpse the intricate lives of microorganisms.

Furthermore, the second review, authored by N. Barros [5], perfectly satisfies the Guest Editors' expectations by introducing an innovative approach for studying soil microbiota. The review presents the state of the art of the very intriguing and promising approach of characterizing the thermodynamics of soil microbial metabolism as a potential tool for the in-depth assessment and comprehension of their strategies for survival, as well as defining their evolutionary state. The author pinpoints the fact that the still unexplored role of microbial diversity—using the energy from the soil organic substrates, and, therefore, the who, where, with whom, and why of managing that energy—could be assessed by unraveling the nature of the soil organic substrates and by monitoring the energy released by the soil microbial metabolism (decomposition vs. assimilation of soil organic substrates). Moreover, the author is right that soil organic content/matter needs differentiation in order to explain the soil carbon cycle in a more appropriate, meaningful and detailed manner.

## 3. Conclusions and Outlook

There is still a long way to go, and a lot of research to do, so as to adequately—if it is possible at all (?)—answer the fundamental questions posed in the present special issue. As partially shown in our special issue (recent case studies and reviews), molecular microbial ecology, i.e., its available tools and those continuously evolving, no doubt has the potential to further enlighten the (still) black box, soil. In this regard, e.g., the discriminatory study of the extracellular (exDNA) and intracellular (iDNA) fractions of the total environmental DNA pool (eDNA), might be a promising approach to (i) further increase the overall information stored about microbiota in soil, e.g., [6], and other environments [7], including specific habitats such as deadwood [8], and (ii) to correctly interpret, critically discuss, and draw relevant conclusions of DNA-based results. Moreover, in the era of culture-independent high-throughput molecular analyses coupled with advanced ecological networking via bioinformatics, the basic, defining steps of any experiment must be taken seriously and correctly, i.e., the experimental design including sampling strategy, soil storage and DNA extraction methods (reviewed by [9,10]). Furthermore, culture-dependent methods (classical microbiology) must not be neglected, and efforts have to be made to increase the number of cultivable microorganisms for further characterization. In fact, the combination of culture-dependent and culture-independent analyses of soil microbiota, and that of all other ecosystems, including aquatic systems, is recommended now more than ever, and can be considered key to the in-depth answer to the question: *Who is Where with Whom and Why?*

**Author Contributions:** J.A.-J. and M.P.: writing of the editorial and supervision of the special issue. All authors have read and agreed to the published version of the manuscript.

**Funding:** This research received no external funding.

**Acknowledgments:** This special issue would not be possible without the contributions of passionate, enthusiastic scientists acting as authors or reviewers, and the dedicated editorial team of Applied Sciences; this is now the best opportunity to express our sincere gratitude and thank all the protagonists! Congratulations to all authors (!) and special thanks to the Section Managing Editor and all involved Editors of Applied Sciences, for their great support and guidance during the adventure of this special issue, and in general, for giving us the opportunity to act as Guest-Editors.

**Conflicts of Interest:** The authors declare no conflict of interest.

## References

1. Zottele, M.; Mayerhofer, J.; Embleton, H.; Wechselberger, K.; Enkerli, J.; Strasser, H. Biological *Diabrotica* Management and Monitoring of *Metarhizium* Diversity in Austrian Maize Fields Following Mass Application of the Entomopathogen *Metarhizium brunneum*. *Appl. Sci.* **2021**, *11*, 9445. [CrossRef]
2. Veresoglou, S.D.; Grünfeld, L.; Mola, M. Micro-Landscape Dependent Changes in Arbuscular Mycorrhizal Fungal Community Structure. *Appl. Sci.* **2021**, *11*, 5297. [CrossRef]
3. Sansupa, C.; Wahdan, S.F.M.; Hossen, S.; Disayathanoowat, T.; Wubet, T.; Purahong, W. Can We Use Functional Annotation of Prokaryotic Taxa (FAPROTAX) to Assign the Ecological Functions of Soil Bacteria? *Appl. Sci.* **2021**, *11*, 688. [CrossRef]
4. Mandolini, E.; Probst, M.; Peintner, U. Methods for Studying Bacterial–Fungal Interactions in the Microenvironments of Soil. *Appl. Sci.* **2021**, *11*, 9182. [CrossRef]
5. Barros, N. Thermodynamics of Soil Microbial Metabolism: Applications and Functions. *Appl. Sci.* **2021**, *11*, 4962. [CrossRef]

6. Gómez-Brandón, M.; Ascher-Jenull, J.; Bardelli, T.; Fornasier, F.; Sartori, G.; Pietramellara, G.; Arfaioli, P.; Egli, M.; Beylich, A.; Insam, H. Ground cover and slope exposure effects on micro- and mesobiota in forest soils. *Ecol. Ind.* **2017**, *80*, 174–185. [CrossRef]
7. Nagler, M.; Podmirseg, S.M.; Ascher-Jenull, J.; Sint, D.; Traugott, M. Why eDNA fractions need consideration in biomonitoring. *Molec. Ecol. Res.* 2022; *early view*. [CrossRef]
8. Probst, M.; Ascher-Jenull, J.; Insam, H.; Gomez-Brandon, M. The Molecular Information about Deadwood Bacteriomes Partly Depends on the Targeted Environmental DNA. *Front. Microbiol.* **2021**, *12*, 640386. [CrossRef] [PubMed]
9. Nannipieri, P.; Ascher, J.; Ceccherini, M.T.; Landi, L.; Pietramellara, G.; Renella, G. Microbial diversity and soil functions. *Eur. J. Soil Sci.* **2017**, *54*, 655–670, Reprinted in *Eur. J. Soil Sci.* **2017**, *68*, 12–26. [CrossRef]
10. Nannipieri, P.; Ascher-Jenull, J.; Ceccherini, M.T.; Pietramellara, G.; Renella, G.; Schloter, M. Beyond microbial diversity for predicting soil functions. *Pedosphere* **2020**, *30*, 5–17. [CrossRef]

Article

# Biological *Diabrotica* Management and Monitoring of *Metarhizium* Diversity in Austrian Maize Fields Following Mass Application of the Entomopathogen *Metarhizium brunneum*

**Maria Zottele [1,*], Johanna Mayerhofer [2], Hannah Embleton [1], Katharina Wechselberger [3], Jürg Enkerli [2] and Hermann Strasser [1]**

1 Department of Microbiology, Leopold-Franzens University Innsbruck, 6020 Innsbruck, Austria; hannah.embleton@student.uibk.ac.at (H.E.); hermann.strasser@uibk.ac.at (H.S.)
2 Molecular Ecology, Agroscope, 8046 Zurich, Switzerland; johanna.mayerhofer@agroscope.admin.ch (J.M.); juerg.enkerli@agroscope.admin.ch (J.E.)
3 Austrian Agency for Health and Food Safety (AGES), 1220 Vienna, Austria; katharina.wechselberger@ages.at
* Correspondence: maria.zottele@uibk.ac.at

**Abstract:** Inundative mass application of *Metarhizium brunneum* BIPESCO 5 (Hypocreales, Clavicipitaceae) is used for the biological control of *Diabrotica v. virgifera* (Coleoptera, Chrysomelidae). Long-term field trials were performed in three Austrian maize fields—with different cultivation techniques and infestation rates—in order to evaluate the efficacy of the treatment to control the pest larvae. In addition, the indigenous *Metarhizium* spp. population structure was assessed to compare the different field sites with BIPESCO 5 mass application. Annual application of the product Granmet-P™ (*Metarhizium* colonized barley kernels) significantly increased the density of *Metarhizium* spp. in the treated soil above the upper natural background level of 1000 colony forming units per gram dry weight soil. Although a decrease in the pest population over time was not achieved in heavily infested areas, less damage occurred in treated field sites in comparison to control sites. The *Metarhizium* population structure was significantly different between the treated field sites. Results showed that inundative mass application should be repeated regularly to achieve good persistence of the biological control agent, and indicated that despite intensive applications, indigenous populations of *Metarhizium* spp. can coexist in these habitats. To date, crop rotation remains the method of choice for pest reduction in Europe, however continuous and preventive application of *M. brunneum* may also present an alternative for the successful biological control of *Diabrotica*.

**Keywords:** *Metarhizium* spp.; *Diabrotica v. virgifera*; inundative application; abundance; population genetics

**Citation:** Zottele, M.; Mayerhofer, J.; Embleton, H.; Wechselberger, K.; Enkerli, J.; Strasser, H. Biological *Diabrotica* Management and Monitoring of *Metarhizium* Diversity in Austrian Maize Fields Following Mass Application of the Entomopathogen *Metarhizium brunneum*. *Appl. Sci.* **2021**, *11*, 9445. https://doi.org/10.3390/app11209445

Academic Editor: Filomena De Leo

Received: 31 August 2021
Accepted: 8 October 2021
Published: 12 October 2021



## 1. Introduction

The western corn root worm *Diabrotica virgifera virgifera* LeConte (Coleoptera, Chrysomelidae), an accidentally introduced, but now firmly established maize pest, causes major damage in maize growing areas in Austria, particularly in regions of Southeast Styria with continuous maize cultivation. The application of the biocontrol agent *Metarhizium brunneum* (Petch) against *D. v. virgifera* larvae has been investigated in a few studies [1–3], but very limited data is available on long-term field trials using only the entomopathogenic fungus (EPF) as insecticide against the maize pest *Diabrotica*. The inundative use of *Metarhizium* aims at controlling the pest within a short period of time and the application has to be repeated if the pest population increases again, because reproduction and/or a permanent establishment of the fungus is not expected. This strategy is mostly used for short-term crops where high population densities of the pest need to be controlled to prevent damage [4]. Large amounts of the biocontrol agent are necessary to achieve a control effect, as soil dwelling insects can only be infected by direct contact with spores. However, annual field processing such as mechanical cultivating or ploughing bears the risk of substantially diminishing the applied microbial agent [5,6]. It was suggested by Rauch et al. [1], that the fungus should be applied preventively, before the pest has established a large population, and pest pressure is still low (i.e., number of beetles should not exceed the economic threshold value of approximately one beetle per ten plants for continuous maize cultivation).

Soil is an extremely complex milieu, an environment with a high number of diverse microorganisms [7,8]. The presence of a viable soil microbiota have an impact on the persistence and/or efficacy of entomopathogenic fungi and vice versa. Therefore, studies on diversity and distribution of soil inhabitants, especially *Metarhizium*, are requested and further knowledge is needed [9–11]. The *Metarhizium* community is influenced by changes in agricultural practice, e.g., abundance changes depending on the crop [10] as well as on type of land-use [12]. After application in high doses, biocontrol agents are exposed to resource restrictions and compete with the well-established indigenous microorganisms [5] including native *Metarhizium* strains. Mayerhofer et al. [9,13] investigated effects of *M. brunneum*-based control agents on microbial communities in pot and field experiments. They found small effects in some treatments, but these were attributed to the product formulation and not to the activity of the fungus itself. However, knowledge on microbial interactions is still limited. Further studies are necessary to assess on the one hand possible effects of mass application of specific strains on microbial communities [10,13], and on the other hand how microbial communities may affect the establishment and development of applied EPF strains.

In this study, we investigated whether long-term inundative mass application of the biocontrol agent *M. brunneum* BIPESCO 5 improves the efficacy of this EPF to control *D. v. virgifera*. In addition, the persistence of the application strain was evaluated with and without the presence of target pest and in co-occurrence with the indigenous *Metarhizium* species and genotypes. In Styria, the pest pressure was high at the beginning of our investigations in 2012, the fungus was applied annually in a six-year long efficacy study to investigate the long-term control effect of the fungus against the larvae of *Diabrotica*. In addition, a preventive application strategy was tested in Tyrol, by annually applying the product in a maize field in a noninfested area over a period of three years.

The necessary information was gained by isolation and cultivation of *Metarhizium* species from the soil, by molecular genetic analyses of isolates, by evaluation of fungal and pest densities, as well as by evaluation of effects on plant damage caused by larval feeding.

## 2. Materials and Methods

### 2.1. Field Sites and Cultivation

The field trials were performed in two fields in Styria (Bad Radkersburg) referred to as "Styria 1" (46°41′1.9608″ N, 16°1′6.7008″ E) and "Styria 2" (46°42′42.2028″ N, 15°55′51.798″ E) and one field in Tyrol (Oberndorf/St. Johann in Tirol; 47°30′23.1552″ N, 12°23′32.3844″ E) referred to as "Tyrol". Both fields in Styria were 2.5 ha in size and 6.6 km apart; the field in Tyrol was 13.5 ha in size. The soil type in all field sites was either a mixture of loamy sand and loamy silt (Styria1, Tyrol) or loamy silt (Styria 2). The region Bad Radkersburg was known for decade-long continuous maize cropping, before an official regulation on crop rotation went into effect in 2019. This region is therefore heavily infested with *D. v. virgifera* since 2009 [1]. Thus, a natural population of *Diabrotica* could be expected in all experimental fields and no artificial infestation with any stage of *Diabrotica* was carried out on either the trial or control fields. Up to the start of this study *D. v. virgifera* has not infested the region Tyrol. Consequently, preventive biological pest control in Tyrol was also carried out in the absence of *Diabrotica* infestations.

While in the field Styria 1 maize and *Cucurbita pepo* L. *var. styriaca* were grown in rotation (2012–2014, 2016, 2018 maize; 2015, 2017 pumpkin) in fields Styria 2 and Tyrol maize was grown annually. All the fields were prepared according to common agricultural practice before sowing (i.e., seeding rate was 70,000 seeds ha$^{-1}$ in all years. manure and the mineral fertilizer Nitramoncal™ (13.5% ammonium and 13.5% nitrate; Borealis L.A.T, Austria) were used as fertilizing elements in April and May each).

Control field sites in Styria and Tyrol (both approx. 3.5 ha) were in close proximity to the treated field sites (Styria 2 and Tyrol; <1500 m air distance) and cultivated with maize annually. The following maize seed varieties were sown: Pharaonix RZ 480, Pioneer Hi-Breed Services, in Styria until 2016, thereafter Mexini RZ 450, RAGT Saaten Österreich; in Tyrol only ATLETICO RZ 280; KWS Austria Saat GmbH. was used. The seeding rate in Tyrol was 80,000 seeds ha$^{-1}$ in all years. Manure and DAP 18/46 (EuroChem Agro GmbH, Germany, Mannheim) was used for fertilizing in April each year. The herbicide Laudis® + Aspect® Pro (Bayer Agrar Austria, Austria, Vienna) was applied at a maize growth stage of 13–15 according to the BBCH scale. More detailed agronomical information on the field sites in Styria can also be found in the full paper of Rauch et al. [1]. Weather stations located in the neighborhood of the experimental fields in Styria and Tyrol recorded air temperature, precipitation, relative humidity, daily sunshine duration, and global radiation throughout the whole study.

### 2.2. Treatment with M. brunneum

The entomopathogenic fungus *M. brunneum* strain BIPESCO 5 cultivated on barley kernels and commercialized as GranMet-P™ (Agrifutur, Italy, Alfianello) was used for all the treatments. The product, registered according to Article 53 of Regulation No. 1107/2009 of the European parliament and of the council (emergency situations in plant protection) in Austria for *Amphimallon solsitiale* and *Phyllopertha horticola* control since 2006, was applied using a RAUCH fertiliser spreader AXIS LTC [final dosage of 50 kg ha$^{-1}$—corresponding to $2 \times 10^{12}$ colony forming units (CFU) per ha] and ploughed in the soil using a HORSCH Terrano 5 FM cultivator to a final depth of 5–10 cm, before maize was sown. All field sites were treated annually in spring throughout the years 2016–2018. Styria 2 was treated annually since 2012, Styria 1 was treated once less with the product GranMet-P™ in the same period due to inadequacies in operational management in 2014. Control sites remained untreated. The quality of the applied active agent was confirmed each year by assessing spore density, colonization ability, pureness, and strain identity [14].

### 2.3. Assessment of Metarhizium spp. Abundance in the Field

The abundance of the applied fungus in the soil was assessed by analyzing the CFU from pooled soil samples taken with a soil corer three times a year ($n$ = 9, sample size $\geq$ 40 cores ha$^{-1}$, drawn in a Z-shape across the field area). Sampling was done in spring (before GranMet-P™ application), in midsummer and before harvesting. More than 40 cores per ha were taken in a sandglass shape and pooled for each field before analysis. At least one pooled sample was taken each sampling ($n_{total}$ = 90). Those samples were processed after Längle et al. [14]. In short, the samples were sieved, diluted in 0.1% (wt/vol) Tween® 80 and plated out in four parallels on selective Sabouraud−4%—Glucose agar medium. Colonies morphological identified as *Metarhizium* spp. were counted after incubation at 25 °C and 60% RH for two weeks and, based on the results, the CFU per gram soil dry weight (CFU g$^{-1}$ dry weight) were calculated.

### 2.4. Evaluation of the Control Efficacy of the Entomopathogen M. brunneum

Direct efficacy assessment is hardly possible due to the small size and fragile texture of mycosed larvae. Instead, indirect methods were used: assessment of adult *Diabrotica* emerging from soil and plant damage due to larval root feeding. The number of adult beetles emerging from soil was assessed in all fields in 2016 and 2017 using the trap system published by Rauch et al. [15] which covered an area of 1.1 m$^2$, equal to 5 cut down maize plants. Traps were installed in late June with a distance of 15 (Styria) and 35 (Tyrol) rows, respectively. Per row up to eight trap systems were established, corresponding to a final trap number of 50 traps per ha in Styria and 24 traps per ha in Tyrol. Emerging beetles were counted at least once a week until week 34 (Styria) and 37 (Tyrol). Additionally, PAL Pheromon-sticky trap systems (CSALOMON™, 1525 Budapest, Hungary) were installed in Styria and Tyrol to monitor the number of beetles on the field sites in the region.

Plant damage was assessed one week before harvesting (in Styria calendar week 35, in Tyrol cw 37), according to BBCH coding 87–91 [16], after Rauch et al. [1], ranking from completely upright (Class 1) to not harvestable due to lodging (Class 4). More than 3000 plants per site and year were assessed.

### 2.5. Genetic Identification of *Metarhizium* spp. Isolates

Two *Metarhizium* isolates were randomly selected for multilocus genotyping (MLG) from plates used for CFU counting ($n$ = 653). DNA extraction was performed after Kepler et al. [17] using an ACME extraction buffer containing 0.05 g diatomaceous earth per 50 mL extraction buffer. DNA extracts were stored at −20 °C until further use. Simple sequence repeat (SSR) PCR was performed according to Mayerhofer et al. [18], using set I and V of the published SSR marker sets (Ma2049, Ma2054, Ma2063, Ma195, Ma327, and Ma2287). PCR products were examined using an ABI 3130xL (Applied Biosystems, Waltham, MA, USA) and the amplicon sizes determined using the software GeneMarker (SoftGenetics; State College, PA, USA).

For each MLG one isolate was selected to determine species allocation by sequencing the 5′ end of nuclear translation elongation factor-1$\alpha$ (5′-TEF-1$\alpha$) and subsequent sequence alignment with sequences of reference strains as described by Fernandez et al. [12]. The 5′-TEF-1$\alpha$ was PCR amplified using alignment with reference sequences primers EF2F (5′-GGAGGACAAGACTCACAT-CAACG-3′) and EFjmetaR (5′-TGCTCACGRGTCTGGC-CATCCTT-3′). The PCR was performed in volumes of 20 µL containing 3 µL DNA extract, 1× Phusion HF Buffer with 7.5 mM $MgCl_2$, 3% DMSO, 0.2 mM dNTPs, 0.2 µM of each primer, and 0.4 U Phusion Polymerase HotStart II. PCR amplification included an initial denaturation at 98 °C for 30 s, followed 38 cycles of denaturation at 98 °C for 5 s, annealing of the primer at 58 °C for 20 s and elongation at 72 °C for 1 min, and a final elongation at 72 °C for 10 min. Quality of the PCR products was verified by agarose gel electrophoresis. PCR products were purified using a Millipore MultiScreen® 96-well filtration plate (Millipore, Darmstadt, Germany) according to the manufacturer´s protocol. Sequencing was performed with the primers mentioned above using the BigDye® Terminator v3.1 cycle sequencing kit (Applied Biosystems, Waltham, MA, USA). Sequencing products were purified with the XTerminator Purification Kit (Applied Biosystems, Waltham, MA, USA) and analyzed using an ABI 3130xL genetic analyzer. Complimentary sequences were assembled using DNA Baser Assembler v4.36.0 (Heracle BioSoft, Mioveni, Romania). Sequences were aligned with reference sequences obtained from the GenBank database representing the different species of the *M. anisopliae* species complex [19,20] using BioEdit [21], a phylogenetic tree was calculated based on the alignment using MEGA X [22].

### 2.6. Data Analysis

Statistical analyses were performed using IBM SPSS Statistics version 23 (IBM Corporation, Armonk, NY, USA), OriginPro 2015G (OriginLab Corporation, Northampton, MA, USA), and R version 1.4.1717 (Free Software Foundation, Inc., Boston, MA, USA). The influence of treatment and time on CFU was analyzed with ANOVA. The correlation between CFU and percentage of BIPESCO 5, CFU and time and temperature and radial growth was assessed with Pearson correlation calculation. Minimum spanning network was created using the "poppr" package of R. For further analysis (e.g., NMDS, PERMANOVA) the package "vegan" was used. The differences of the *Metarhizium* population structure between locations were assessed with the "adonis" function within the "vegan" package based on abundance of SSR derived multilocus genotypes and Bray–Curtis dissimilarity matrices.

## 3. Results

### 3.1. Evaluation of *Metarhizium* spp. Abundance

The *Metarhizium* spp. abundance in all treated field sites increased after application of the production strain. Although achieved CFU values in field site Styria 2 fluctuated during the seasons between 1480 and 53,850 CFU $g^{-1}$ dry weight soil (Figure 1A), this site, continuously planted with maize and treated annually with GranMet-P$^{TM}$ since 2012, has consistently shown significantly higher *Metarhizium* CFU values than the untreated control site in all samples taken since spring 2015. A weak to moderate positive correlation between CFU and time ($r = 0.4$, $p < 0.001$) was determined for this field site (from first to last sampling). Styria 1 showed the highest variation in *Metarhizium* CFU (Figure 1B; mean values of 720 up to 85,580 CFU $g^{-1}$ dry weight soil after first application). Although high CFU levels were not able to persist (no correlation between CFU and time; $r = 0.06$, $p = 0.61$), development of CFU was significantly different from control site ($p < 0.001$). In all Styrian field sites, including the control, *Metarhizium* CFU increased significantly since first sampling in 2012 ($p < 0.001$)—where less than 100 CFU were found—up to at least 2800 CFU (evaluated in the soil of the control field in 2018).

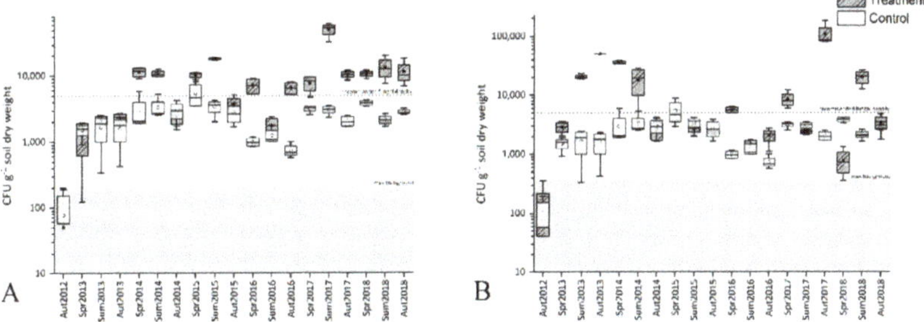


**Figure 1.** *Metarhizium* spp. abundance in GranMet-P$^{TM}$ applied Styrian field sites. (**A**) shows soil samples taken from field site Styria 2, (**B**) Styria 1, both compared to untreated control site. Samples were taken in spring (Spr), summer (Sum), and autumn (Aut) from autumn 2012 to autumn 2018. The grey box indicates maximal background CFU levels before treatment, the dotted lines show recommended fungal density for sustainable control in the soil [1]. Lines and dots in box plots show median and mean CFU $g^{-1}$ soil dry weight, respectively; boxes show the 25th and 75th percentiles, whiskers the 10th and 90th percentiles.

After the first application in Tyrol (March 2016), the abundance of *Metarhizium* spp. increased from a maximum background value of 2500 CFU $g^{-1}$ dry weight soil to densities above the recommended abundance of 5000 CFU $g^{-1}$ dry weight soil (Figure 2). Significantly higher values were achieved after the second treatment and could be established throughout the last year of the field trial with a final fungal density of 11,386 CFU $g^{-1}$ dry weight soil. In comparison to the increase of the *Metarhizium* spp. density in the treated field site ($r = 0.41$; $p < 0.001$), the CFU in the control fields in Tyrol showed a negative correlation of CFU and time ($r = -0.48$; $p = 0.003$). A decrease in CFU by a factor of five could be observed in samples from autumn 2018 ($366 \pm 223$ CFU) compared to the first sampling in 2016 ($1912 \pm 536$ CFU) on this experimental site.

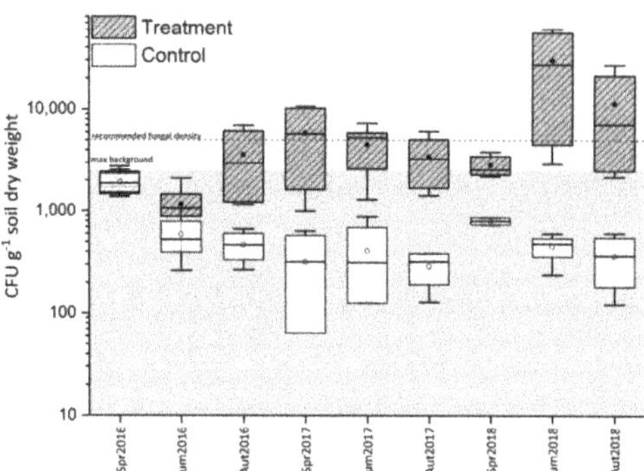


**Figure 2.** *Metarhizium* spp. abundance in Tyrolean field sites. Samples were taken in spring (Spr), summer (Sum), and autumn (Aut) from 2016 to 2018. Grey box indicates maximal background CFU levels before treatment, dotted line shows recommended fungal density. Lines and dots in box plots show median and mean CFU $g^{-1}$ soil dry weight, respectively; boxes show the 25th and 75th percentiles, whiskers the 10th and 90th percentiles.

A significant positive correlation between percentage of isolates identified as BIPESCO 5 and amount of CFU was assessed for Styria 1 ($r = 0.996$; $p < 0.001$) and Tyrol ($r = 0.742$; $p = 0.03$). A moderate, but non-significant correlation occurred at the field site Styria 2 ($r = 0.527$; $p = 0.18$). The sustainable establishment of BIPESCO 5 varied between the field sites: Styria 1, which showed the highest variability of *Metarhizium* spp. abundance, also showed poor persistence of BIPESCO 5 over time, while in all the other treated fields BIPESCO 5 was able to persist throughout the season.

### 3.2. Pest Abundance and Plant Injury

In the heavily infested areas in Styria the *D. v. virgifera* population density continued to increase over the years. On average, the number of caught beetles doubled every year, up to 130 beetles per $m^2$ caught with the emergence trap system in 2018 on the untreated maize field. Only the crop rotation in combination with the biocontrol agent in Styria 1 ensured a significant reduction of the pest, with only five emerging beetles per $m^2$. The number of adult *Diabrotica* evaluated in Styria was significantly different between all field sites ($p < 0.001$, data not shown). Although *Diabrotica* population pressure in Styrian maize fields was very high (Table 1), both treated field sites showed no or only low damage of maize plants. As for the untreated control area, the extent of the damage was affected by the prevailing weather conditions. In the field season 2016 more than 30% of the maize plants showed plant lodging. In 2017, no lodging was observed, but due to the lack of water, plants dried up, were low growing, and fewer-to-no corn cobs had developed. In 2018, less than 1.25% of plants showed lodging due to the sufficient precipitation during this season (Figure 3). Overall, no root injuries were observed in 2018.

**Table 1.** Number of beetles in the trial region evaluated with PAL sticky traps (CSALOMON$^{TM}$, Hungary) in the regions Bad Radkersburg and Oberndorf/St. Johann in Tyrol from 2016 to 2020. Shown is the mean number of beetles over the season per trap with minimum (min) and maximum (max) number of beetles per week.

| | Bad Radkersburg (Styria) | | | | | | Oberndorf/ St. Johann I. T. (Tyrol) | | | | | |
|---|---|---|---|---|---|---|---|---|---|---|---|---|
| Year | Mean N° beetles total | Min beetles per week | Max beetles per week | Increase to previous year | cw of monitoring | cw of max catching | Mean N° beetles total | Minbeetles per week | Max beetles per week | Increase to previous year | cw of monitoring | cw of max catching |
| 2016 | x | <250 | >250 | - | 27/41 | - | 260 | 0 | 226 | - | 27/38 | 36 |
| 2017 | 4429 | 0 | 980 | - | 27/38 | 36 | 749 | 0 | 549 | 2.88 | 27/37 | 31 |
| 2018 | 7336 | 0 | 1488 | 1.66 | 27/38 | 36 | 1008 | 0 | 475 | 1.35 | 27/38 | 32 |
| 2019 | 4588 | 10 | 1041 | 0.63 | 27/38 | 36 | 753 | 0 | 250 | 0.75 | 28/40 | 37 |
| 2020 | 7023 | 72 | 1193 | 1.53 | 27/38 | 37 | y | - | - | - | - | - |

cw calendar week; x the exact number of beetles was not evaluated, only classified as <250 or >250 beetles per trap; y not evaluated;—not calculated due to missing data.

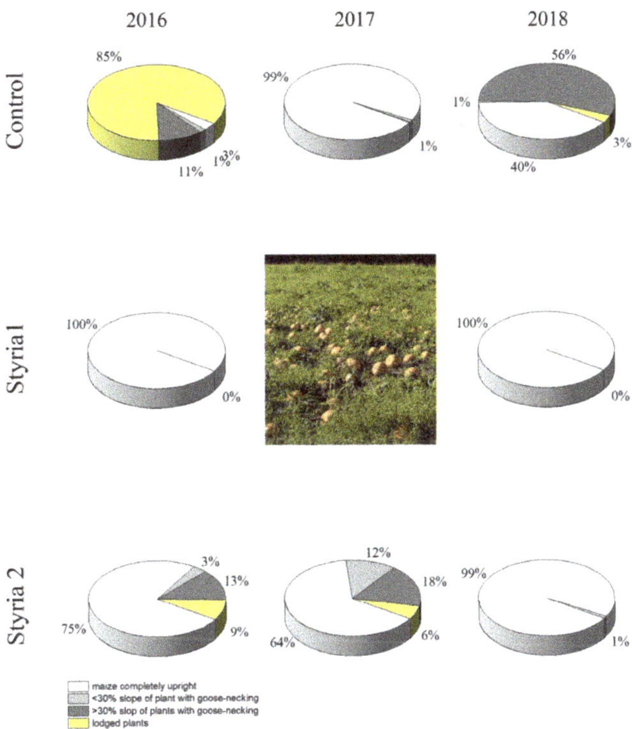


**Figure 3.** Percentage of plants for each damage level assessed in the years 2016 to 2018 in Styria. Healthy plants are indicated as upright, completely fallen plants as lodging. The damage for Styria 1 in 2017 could not be surveyed due to the cultivation of pumpkin (*Cucurbita pepo* L. *var. styriaca*) as part of the crop rotation.

In Tyrol, a total of only two beetles were caught in the emerging trap, confirming that the pest has reached Tyrol, however, in such small numbers that damage to the crop was not to be expected: plant health was not yet affected by larval root feeding, all plants were scored class 1 according to Rauch et al. [1]. Nevertheless, data obtained from the pheromone traps revealed a two-to-three-fold increase in *Diabrotica* density per year (Table 1).

### 3.3. *Metarhizium* Genotyping

SSR-based genotyping and subsequent 5′-TEF-1α sequencing of 653 *Metarhizium* isolates revealed the presence of 31 multilocus genotypes (MLGs) in addition to the applied production strain (Figure 4). The MLGs represented three species, i.e., *M. brunneum*, *M. robertsii*, and *M. lepidiotae* (Table 2; Figure S1).

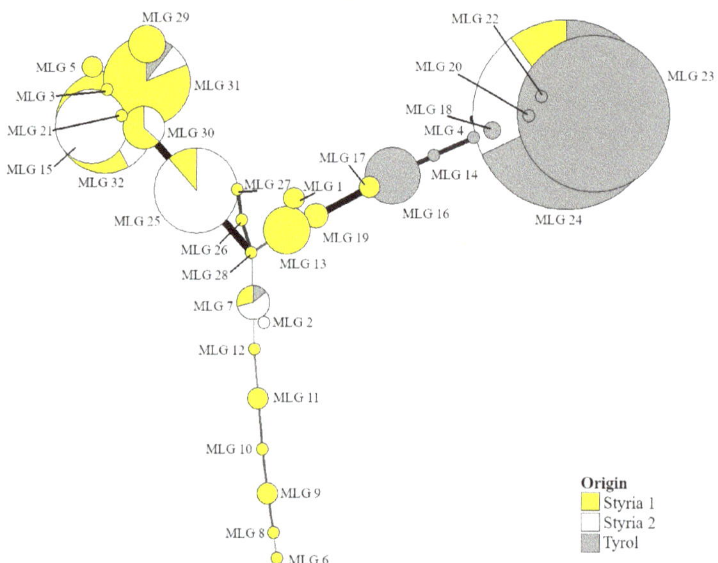


**Figure 4.** Minimum spanning network (MSN) showing the relationship between the SSR genotypes isolated from the treated field sites (Styria 1, Styria 2, and Tyrol). Circle sizes are proportional to the number of isolates belonging to one MLG, the thickness of the line is proportional to genetic SSR based similarity of genotypes. MLG 24 represents the genotype of the applied strain BIPESCO 5.

**Table 2.** Numbers of isolates and MLG identified as *M. brunneum*, *M. robertsii*, or *M. lepidiotae* from the treated field sites. The applied strain BIPESCO 5 is shown separated from *M. brunneum*.

| | | **BIPESCO 5** | ***M. Brunneum*** | | | ***M. Robertsii*** | | | ***M. Lepidiotae*** | | |
|---|---|---|---|---|---|---|---|---|---|---|---|
| **Origin** | **Year** | **N** | **N** | **MLG** | **SG** | **N** | **MLG** | **SG** | **N** | **MLG** | **SG** |
| Styria 1 | 2016 | 0 | 25 | 1 | 0 | 47 | 7 | 2 | 2 | 2 | 2 |
| Styria 1 | 2017 | 15 | 10 | 1 | 0 | 34 | 8 | 3 | 13 | 8 | 5 |
| Styria 1 | 2018 | 0 | 8 | 2 | 0 | 16 | 3 | 2 | 1 | 1 | 1 |
| Styria 2 | 2016 | 11 | 28 | 2 | 0 | 33 | 4 | 0 | 0 | 0 | 0 |
| Styria 2 | 2017 | 24 | 29 | 2 | 0 | 19 | 4 | 3 | 0 | 0 | 0 |
| Styria 2 | 2018 | 12 | 3 | 1 | 0 | 11 | 1 | 1 | 0 | 0 | 0 |
| Tyrol | 2016 | 20 | 130 | 5 | 3 | 5 | 1 | 0 | 0 | 0 | 0 |
| Tyrol | 2017 | 101 | 42 | 5 | 3 | 1 | 1 | 1 | 0 | 0 | 0 |
| Tyrol | 2018 | 30 | 2 | 2 | 2 | 0 | 0 | 0 | 0 | 0 | 0 |

N total number of isolates; MLG number of unique multilocus genotypes; SG MLG with a single isolate.

The MLG of the applied strain *M. brunneum* BIPESCO 5 was detected in 213 isolates (32.6%) and at least once in every treated field site after application of the product. The MLG composition without the applied strain was significantly different (PERMANOVA, $p < 0.001$) among the three locations (Figure 5). *M. brunneum* and *M. robertsii* were isolated from all field sites. *M. robertsii* was the dominant (54.3% in Styria 1, 67% in Styria 2) and, for Styria 2, genetically most diverse species in both Styrian trial sites, excluding the applied strain. Field site Styria 1 contained the highest diversity of MLGs from the three analyzed locations. Fifty-three percent of all MLGs were only found there. It was also the only field site where the species *M. lepidiotae* occurred (37.5% of the genotypes; but only corresponding to 9.9% of all isolates—the individual MLGs were usually found once). In Tyrol, *M. brunneum* was the dominating species isolated from the soil (96.7% of samples without BIPESCO 5), exhibiting one major genotype (82.1% of all isolates). Two genotypes (MLG 7, MLG 31) were found in all three sampling sites. The majority (22) of the 31 MLGs was only isolated in Styria. Most of the MLGs (84%) were only found in one of the field sites. None of the genotypes was present at every sampling point of the different sites, but in both Styrian fields two genotypes were isolated at least 75% of the sampling times (Styria 1: MLG 31, MLG 32; Styria 2: MLG 15, MLG 32). In Tyrol, only one MLG was found in 75% of the samplings (MLG 23), but another genotype was found at least at 62.5% of the sampling times (MLG 16). Out of ten different *M. brunneum* genotypes, only four clustered closely to the applied strain BIPESCO 5 (Figure 4; MLG 18, 20, 22, and 23). A maximum of two SSR loci differed in these genotypes by a maximum of four base pairs. The other, non-clustering *M. brunneum* genotypes contained at least five variable loci (out of six).

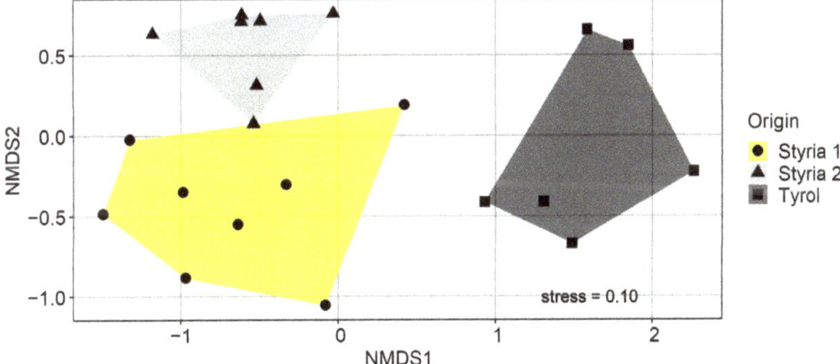


**Figure 5.** Non-metric multidimensional scaling (NMDS) using abundance of MLGs based on Bray–Curtis dissimilarities showing a different *Metarhizium* population structure at the treated field sites Styria 1 (circles), Styria 2 (triangle), and Tyrol (square). Each symbol represents the population composition at a sampling time. Data shown is without the applied strain and from samples after the first application of the product. Hulls illustrate the different field sites Styria 1 (yellow), Styria 2 (light grey), and Tyrol (dark grey).

## 4. Discussion

*Diabrotica virgifera virgifera* has become one of the most important maize pests in Europe and many different studies have been carried out to obtain information on the biology and behavior of the pest, as well as control options against it [1,23–25]. In this study, efficacy of the entomopathogenic fungus *M. brunneum* BIPESCO 5 against *Diabrotica* was compared under the following conditions: the product GranMet-P$^{TM}$ (registered for *Amphimallon solstitiale* and *Phyllopertha horticola* control) was applied at the time of the general tillage in March/April with traditional agricultural equipment, reasonable workload for the farmers, heavily *Diabrotica* infested fields with continuous maize cultivation or regular crop rotation. In addition, continuously cultivated maize fields not yet heavily infested were treated to ensure the establishment of the entomopathogenic fungus without the presence of the target pests as a preventive measure by increasing the pest antagonist.

We found significant differences in emerging adults in all heavily infested areas. The lowest number of adult beetles was observed in the field site with crop rotation ($\geq$5 adults m$^{-2}$). This low number of adults per m$^2$ compared to an at least six times higher number found in continuous maize fields (treatment and control) is in accordance with results reported by Szalai et al. [26]. Although oviposition into non-maize fields near heavily infested maize fields occurs and therefore adult emergence in first-year maize can be observed, crop rotation still is the most effective method to quickly decrease *Diabrotica* population in maize fields in Europe [26–28]. As already reported by Rauch et al. [1], *M. brunneum* alone was not able to reduce the *Diabrotica* population below an acceptable/zero-damage threshold level in our study due to the high pest population density (i.e., economic threshold value: >1 beetle per plant during any weekly counts in July and August, [29]). Nevertheless, plant health was better, and less lodging occurred in *Metarhizium* treated field sites. The beneficial effect of *Metarhizium* on different plants was also recognized in studies on rhizosphere colonization of the entomopathogenic fungus; results showed extensive root development, increased root length, improved plant growth, decreased stress in plants and improved availability of nutrients [30–32]. For maize crops in particular, it was shown, that, for instance, plant-growth-promoters were activated by the production of auxins at the roots by *Metarhizium* spp. [31]. Furthermore, entomopathogenic fungi could colonize niches which otherwise are occupied by plant pathogens [31,33].

Persistence of *M. brunneum* at elevated abundance of approximately 5000 CFU g$^{-1}$ dry weight in soil is important for the successful control of soil-borne pests such as *Diabrotica* [1]. In Tyrol and Styria 2, BIPESCO 5 could be established (Table 2) and persisted in this density. The strain also persisted in the field site Styria 1, but not as sustainably as in the other field sites—here, annual reapplication was necessary to ensure the persistence of the strain throughout the planting season.

Investigations on the diversity of *Metarhizium* in soil has revealed a variable distribution of the different species worldwide, but with genetically closely related isolates across large distances [34]. Klingen et al. [35] also found higher diversity of entomopathogenic fungi in organically farmed soil compared to conventionally treated field sites. Liao et al. [31] reported that there is evidence that plant host associations play an important role in evolutionary divergence within the genus *Metarhizium*. In the USA, *M. brunneum* is associated with the rhizosphere of shrubs and trees, whereas *M. robertsii* is found in open fields and grassland [10,36]. In addition, *M. brunneum* was only found in agricultural and open field sites when *M. robertsii* was also present [36]. In contrast, *M. brunneum* has been reported to be the dominant species in agricultural and grassland fields in Europe [12,37]. This could also be observed in our field sites in Tyrol, with more than 96.7% of all isolates being *M. brunneum* (without the applied strain). Although a comparison of the distribution of *Metarhizium* species already published is difficult due to different sampling techniques, variable distribution patterns of dominant species were found in recent studies. The species *M. anisopliae*, *M. brunneum*, *M. robertsii*, and *M. pingshaense* are most frequently isolated [38], and even these species are found with local preferences: *M. pingshaense* being the most common species found in soil taken from various locations in Japan [39], *M. anisopliae* in agricultural soil in Brazil [40,41], and *M. robertsii* dominating in the USA [10], for example. Other species only occur in restricted areas. *M. flavoviride*, for example, was the most common species found in agricultural fields in Denmark [42]. These global distribution differences of *Metarhizium* species can also be found on a smaller scale in our test sites—which all had a significantly different *Metarhizium* population structure. The two chosen regions differ in their climate, landscape and vegetation—the wide-open space in Southeast Styria, 209 m above (Adriatic) sea level (Pannonic climate), and the mountainous region of North-Tyrol (659 m above sea level; Alpine climate). Overall, *M. brunneum* as well as *M. robertsii* and *M. lepidiotae* were isolated as indigenous species. At the sampling sites of Tyrol, most of the isolated colonies were classified as *M. brunneum*, while in Styria, *M. robertsii* dominated the *Metarhizium* community. Bidochka et al. [43] suggested that due to the saprophytic phase of the species, their population genetics shifted in accordance with the ability to grow at low or high temperatures. This would correspond to our preliminary studies on the radial growth of the three species (Table S1) in which growth between 25 °C and 30 °C for *M. brunneum* (negative correlation of temperatures with $r = -0.559$, $p < 0.05$) and *M. robertsii* ($r = -0.018$, $p > 0.05$) stagnates or declines and *M. lepidiotae*, which only occurs in the warmer region of Styria, showing a larger colony diameter at 30 °C compared to incubation at 25 °C ($r = 0.641$, $p < 0.05$). These findings are supported by Kryukov et al. [44], who have reported, that *M. robertsii* and *M. brunneum* have different optimal growth temperatures, with *M. robertsii* preferring higher temperatures than *M. brunneum*. In contrast Steinwender et al. [37] have found that certain *Metarhizium* species are not necessarily dominant in sun exposed habitats but react differently to multiple abiotic factors. Regarding *Metarhizium* spp. abundance it is well documented that there is a high correlation between temperature, humidity and survival of entomopathogenic fungi [45]. These natural abiotic factors may have an influence on the survival and development of the *M. brunneum* production strain in both Austrian regions. Our data suggest that *M. brunneum* BIPESCO 5 is more persistent when applied in Tyrol. Meyling and Eilenberg [46] summarized that *Metarhizium* is more common in exposed and regularly disturbed soil but cannot extensively proliferate. In addition, different tillage systems lead to very variable soil densities. This may also be the cause for the fluctuation of CFU and genotypes isolated from soil of the sampling site Styria 1: due to the crop rotation applied, farming practices were different compared to the field sites in Tyrol and Styria 2, where, for instance, the tillage practice remained the same every year.

Crop rotation remains the option of choice for rapid pest population decline at high pest densities. However, both the preventive and continuous use of GranMet-P$^{TM}$ can successfully increase the density of the entomopathogenic fungus in the soil, and therefore may decline the pest population in the regions. In addition, healthy and vigorous plant growth is promoted. The production strain of the GranMet-P$^{TM}$ product—BIPESCO 5—has been successfully tested for western corn rootworm control in Austria, Germany, Hungary, Italy and Switzerland [1,3,25,47,48]. However, studies on the biological control of adult beetles and marketable products are still lacking, although preliminary studies [49] have shown promising results. Further findings in this area will contribute to an even greater success of biological control of *Diabrotica* populations.

**Supplementary Materials:** The following are available online at https://www.mdpi.com/article/10.3390/app11209445/s1, Figure S1: Maximum likelihood phylogenetic tree based on the alignment of 5′-TEF-1α sequences representing 31 different multilocus genotypes (MLG) isolated from the soil (MLG 24 representing the applied strain BIPESCO 5) and reference strains with a total of 672 positions in the final dataset, Table S1: Radial growth (mm) of production strain BIPESCO 5 and isolates from Styria after 14d incubation.

**Author Contributions:** M.Z., H.S., K.W., and J.E. conceived and designed research; M.Z., H.S., and J.M. performed trials and experiments; M.Z., J.M., and H.E. performed data analysis; M.Z., H.S., and H.E. wrote original draft; all authors reviewed and edited the manuscript. All authors have read and agreed to the published version of the manuscript.

**Funding:** This research was funded by the Austrian Federal Ministry for Agriculture, Regions and Tourism and all state governments, DIACONT grant number 10111/2, the European Commission, INBIOSOIL grant number 282767, and supported by a doctoral scholarship of the Leopold-Franzens University Innsbruck (APC included).

**Institutional Review Board Statement:** Not applicable.

**Informed Consent Statement:** Not applicable.

**Data Availability Statement:** All additional data can be obtained from the corresponding author upon request. Raw sequences were deposited in the NCBI BankIt database with the accession numbers OK423535-OK423563.

**Acknowledgments:** The authors would like to thank the farmers Markus Sammer, Fritz Waltl, DI Leo Ladenhauf, Franz Hödl, and Klaus Uidl for providing the experimental and control areas and for their active cooperation as well as Alexandra Gruber and Jaka Zubanič for their assistance on the field and in the laboratory. Thanks also to the companies Agrifutur s.r.l. and Samen Schwarzenberger (Völs, Tirol) for providing GranMet-P$^{TM}$, and Saatbau Linz for providing maize seeds.

**Conflicts of Interest:** The authors declare no conflict of interest.

## References

1. Rauch, H.; Steinwender, B.M.; Mayerhofer, J.; Sigsgaard, L.; Eilenberg, J.; Enkerli, J.; Zelger, R.; Strasser, H. Field efficacy of *Heterorhabditis bacteriophora* (Nematoda: Heterorhabditidae), *Metarhizium brunneum* (Hypocreales: Clavicipitaceae), and chemical insecticide combinations for *Diabrotica virgifera virgifera* larval management. *Biol. Control* **2017**, *107*, 1–10. [CrossRef]
2. Petzold-Maxwell, J.L.; Jaronski, S.T.; Clifton, E.H.; Dunbar, M.W.; Jackson, M.A.; Gassmann, A.J. Interactions among Bt maize, entomopathogens, and rootworm species (Coleoptera: Chrysomelidae) in the field: Effects on survival, yield, and root injury. *J. Econ. Entomol.* **2013**, *106*, 622–632. [CrossRef]
3. Meissle, M.; Pilz, C.; Romeis, J. Susceptibility of *Diabrotica virgifera virgifera* (Coleoptera: Chrysomelidae) to the entomopathogenic fungus *Metarhizium anisopliae* when feeding on *Bacillus thuringiensis* Cry3Bb1-expressing maize. *Appl. Environ. Microbiol.* **2009**, *75*, 3937–3943. [CrossRef]
4. Hajek, A.; Eilenberg, J. Augmentation: Inundative and inoculative biological control. In *Natural Enemies: An Introduction to Biological Control*, 2nd ed.; Cambridge University Press: Cambridge, UK, 2018; pp. 66–84.
5. Köhl, J.; Booij, K.; Kolnaar, R.; Ravensberg, W.J. Ecological arguments to reconsider data requirements regarding the environmental fate of microbial biocontrol agents in the registration procedure in the European Union. *BioControl* **2019**, *64*, 469–487. [CrossRef]
6. Jaronski, S.T. Ecological factors in the inundative use of fungal entomopathogens. *BioControl* **2010**, *55*, 159–185. [CrossRef]
7. Inglis, G.D.; Goettel, M.S.; Butt, T.M.; Strasser, H. Use of hyphomycetous fungi for managing insect pest. In *Fungi as Bicontrol Agents: Progress, Problems and Potential*; Butt, T.M., Jackson, C., Magan, N., Eds.; CABI: Wallingford, UK, 2001; pp. 23–69.
8. Scheepmaker, J.W.A.; Butt, T.M. Natural and released inoculum levels of entomopathogenic fungal biocontrol agents in soil in relation to risk assessment and in accordance with EU regulations. *Biocontrol Sci. Technol.* **2010**, *20*, 503–552. [CrossRef]
9. Mayerhofer, J.; Eckard, S.; Hartmann, M.; Grabenweger, G.; Widmer, F.; Leuchtmann, A.; Enkerli, J. Assessing effects of the entomopathogenic fungus *Metarhizium brunneum* on soil microbial communities in *Agriotes* spp. biological pest control. *FEMS Microbiol. Ecol.* **2017**, *93*, 1–15. [CrossRef]
10. Kepler, R.M.; Ugine, T.A.; Maul, J.E.; Cavigelli, M.A.; Rehner, S.A. Community composition and population genetics of insect pathogenic fungi in the genus *Metarhizium* from soils of a long-term agricultural research system. *Environ. Microbiol.* **2015**, *17*, 2791–2804. [CrossRef]
11. Glare, T.R.; Scholte Op Reimer, Y.; Cummings, N.; Rivas-Franco, F.; Nelson, T.L.; Zimmermann, G. Diversity of the insect pathogenic fungi in the genus *Metarhizium* in New Zealand. *New Zealand J. Bot.* **2021**, *59*, 440–456. [CrossRef]
12. Fernández-Bravo, M.; Gschwend, F.; Mayerhofer, J.; Hug, A.; Widmer, F.; Enkerli, J. Land-use type drives soil population structures of the entomopathogenic fungal genus *Metarhizium*. *Microorg.* **2021**, *9*, 1380. [CrossRef]
13. Mayerhofer, J.; Rauch, H.; Hartmann, M.; Widmer, F.; Gschwend, F.; Strasser, H.; Leuchtmann, A.; Enkerli, J. Response of soil microbial communities to the application of a formulated *Metarhizium brunneum* biocontrol strain. *Biocontrol Sci. Technol.* **2019**, *29*, 547–564. [CrossRef]
14. Längle, T.; Pernfuss, B.; Seger, C.; Strasser, H. Field efficacy evaluation of *Beauveria brongniartii* against *Melolontha melolontha* in potato cultures. *Sydowia* **2005**, *57*, 54–93.
15. Rauch, H.; Zelger, R.; Strasser, H. Highly efficient field emergence trap for quantitative adult western corn rootworm monitoring. *J. Kans. Entomol. Soc.* **2016**, *89*, 256–266. [CrossRef]
16. Meier, U. Growth stages of mono and dicotyledonous plants. In *BBCH monograph*; Federal Biological Research Centre for Agriculture and Forestry: Bonn, Germany, 2001.
17. Kepler, R.M.; Chen, Y.; Kilcrease, J.; Shao, J.; Rehner, S.A. Independent origins of diploidy in *Metarhizium*. *Mycologia* **2014**, *108*, 1091–1103.
18. Mayerhofer, J.; Lutz, A.; Widmer, F.; Rehner, S.A.; Leuchtmann, A.; Enkerli, J. Multiplexed microsatellite markers for seven *Metarhizium* species. *J. Invertebr. Pathol.* **2015**, *132*, 132–134. [CrossRef]
19. Bischoff, J.F.; Rehner, S.A.; Humber, R.A. A multilocus phylogeny of the *Metarhizium anisopliae* lineage. *Mycologia* **2009**, *101*, 512–530. [CrossRef]
20. Rehner, S.A. Genetic structure of *Metarhizium* species in western USA: Finite populations composed of divergent clonal lineages with limited evidence for recent recombination. *J. Invertebr. Pathol.* **2020**, *177*, 107491. [CrossRef]
21. Hall, T.A. BioEdit: A user-friendly biological sequence alignment editor and analysis program for Windows 95/98/NT. *Nucleic Acids Symp. Ser.* **1999**, *41*, 95–98.
22. Kumar, V.; Kakkar, G.; Seal, D.R.; McKenzie, C.L.; Osborne, L.S. Evaluation of insecticides for curative, preventive, and rotational use on *Scirtothrips dorsalis* Sout Asia 1 (Thysanoptera: Thripidae). *Fla Entomol.* **2018**, *100*, 634–646. [CrossRef]
23. Pilz, C.; Keller, S.; Kuhlmann, U.; Töpfer, S. Comparative efficacy assessment of fungi, nematodes and insecticides to control western corn rootworm larvae in maize. *BioControl* **2009**, *54*, 671–684. [CrossRef]

24. Töpfer, S.; Haye, T.; Erlandson, M.; Goettel, M.; Lundgren, J.; Kleespies, R.G.; Weber, D.; Jackson, J.; Peters, A.; Cabrera Walsh, G.; et al. A review of the natural enemies of beetles in the subtribe *Diabroticina* (Coleoptera: Chrysomelidaer): Implications for sustainable pest management. *Biocontrol. Sci. Technol.* **2009**, *19*, 1–65. [CrossRef]
25. Benjamin, E.O.; Grabenweger, G.; Strasser, H.; Wesseler, J. The socioeconomic benefits of biological control of western corn rootworm *Diabrotica virgifera virgifera* and wireworms *Agriotes* spp. in maize and potatoes for selected European countries. *J. Plant Dis. Prot.* **2018**, *125*, 273–285. [CrossRef]
26. Szalai, M.; Papp Komáromi, J.; Bazok, R.; Igrc Barcic, J.; Kiss, J.; Töpfer, S. Generational growth rate estimates of *Diabrotica virgifera virgifera* populations (Coleoptera: Chrysomelidae). *J. Pest Sci.* **2011**, *84*, 133–142. [CrossRef]
27. Töpfer, S.; Zellner, M.; Kuhlmann, U. Food and oviposition preferences of *Diabrotica v. virgifera* in multiple-choice crop habitat situations. *Entomologia* **2013**, *1*, 60–68. [CrossRef]
28. Sivčev, J.; Kljajić, P.; Kostić, M.; Sivčev, L.; Stanković, S. Management of western corn rootworm (*Diabrotica virgifera virgifera*). *Pestic. Phytomed.* **2012**, *27*, 189–201. [CrossRef]
29. Stamm, D.E.; Mayo, Z.B.; Campbell, J.B.; Witkowki, J.F.; Andersen, L.W.; Kozub, R. Western corn rootworm (Coleoptera, Chrysomelidae) beetle counts as a means of making larval control recommendations in Nebraska. *J. Econ. Entomol.* **1985**, *78*, 794–798. [CrossRef]
30. Jaber, L.R.; Enkerli, J. Effect of seed treatment duration on growth and colonization of *Vicia faba* by endophytic *Beauveria bassiana* and *Metarhizium brunneum*. *Biol. Control* **2016**, *103*, 187–195. [CrossRef]
31. Liao, X.; O' Brien, T.R.; Fang, W.; St. Leger, R.J. The plant beneficial effects of *Metarhizium* species correlate with their association with roots. *Appl. Microbiol. Biotechnol.* **2014**, *98*, 7089–7096. [CrossRef]
32. Ahmad, I.; Jiménez-Gasco, M.; Luthe, D.S.; Shakeel, S.N.; Barbercheck, M.E. Endophytic *Metarhizium robertsii* promotes maize growth, supresses insect growth, and alters plant defense gene expression. *Biol. Control* **2020**, *144*, 104167. [CrossRef]
33. Hu, S.; Bidochka, M.J. Root colonization by endophytic insect pathogenic fungi. *J. Appl. Microbiol.* **2021**, *130*, 570–581. [CrossRef]
34. Bidochka, M.; Small, C. Phylogeopgraphy of *Metarhizium*, an insect pathogenic fungus. In *Insect-Fungal Associations: Ecology and Evolution*; Vega, F.E., Blackwell, M., Eds.; Oxford University Press: Oxford, UK, 2005; pp. 28–50.
35. Klingen, I.; Eilenberg, J.; Meadow, R. Effects of farming system, field margins and bait insect on the occurrence of insect pathogenic fungi in soils. *Agric. Ecosyst. Environ.* **2002**, *91*, 191–198. [CrossRef]
36. Wyrebek, M.; Huber, C.; Sasan, R.K.; Bidochka, M.J. Three sympatrically occurring species of *Metarhizium* show plant rhizosphere specificity. *Microbiology* **2011**, *157*, 2904–2911. [CrossRef] [PubMed]
37. Steinwender, B.M.; Enkerli, J.; Widmer, F.; Eilenberg, J.; Thorup- Kristensen, K.; Meyling, N.V. Molecular diversity of the entomopathogenic fungal *Metarhizium* community within an agroecosystem. *J. Invertebr. Pathol.* **2014**, *123*, 6–12. [CrossRef] [PubMed]
38. Rehner, S.A.; Kepler, R.M. Species limits, phylogeography and reproductive mode in the *Metarhizium anisopliae* complex. *J. Invertebr. Pathol.* **2017**, *148*, 60–66. [CrossRef]
39. Nishi, O.; Hasegawa, K.; Iiyama, K.; Yasunaga-Aoki, C.; Shimizu, S. Phylogenetic analysis of *Metarhizium* spp. isolated from soil in Japan. *Appl. Entomol. Zool.* **2011**, *46*, 301–309. [CrossRef]
40. Rocha, L.F.; Inglis, P.W.; Humber, R.A.; Kipnis, A.; Luz, C. Occurrence of *Metarhizium* spp. in central Brazilian soils. *J. Basic Microbiol.* **2013**, *53*, 251–259. [CrossRef]
41. Rezende, J.M.; Zanardo, A.B.R.; da Silva Lopes, M.; Delalibera, I.; Rehner, S.A. Phylogenetic diversity of Brazilian *Metarhizium* associated with sugarcane agriculture. *BioControl* **2015**, *60*, 495–505. [CrossRef]
42. Keyser, C.A.; de Fine, H.H.; Steinwender, B.M.; Meyling, N.V. Diversity within the entomopathogenic fungal species *Metarhizium flavoviride* associated with agricultural crops in Denmark. *BMC Microbiol.* **2015**, *15*, 249. [CrossRef]
43. Bidochka, M.; Kamp, A.M.; Lavender, T.M.; Dekoning, J.; de Croos, J.N.A. Habitat association in two genetic groups in the insect-pathogenic fungus *Metarhizium anisopliae*: Uncovering cryptic species? *Appl. Environ. Microbiol.* **2001**, *67*, 1335–1342. [CrossRef]
44. Kryukov, V.; Yaroslavtseva, O.; Tyurin, M.; Akhanaev, Y.; Elisaphenko, E.; Wen, T.C.; Tomilova, O.; Tokarev, Y.; Glupov, V. Ecological preferences of *Metarhizium* spp. from Russia and neighboring territories and their activity against Colorado potato beetle larvae. *J. Invertebr. Pathol.* **2017**, *149*, 1–7. [CrossRef]
45. Daoust, R.A.; Roberts, D.W. Studies on the prolonged storage of *Metarhizium anisopliae* conidia: Effect of temperature and relative humidity on conidial viability and virulence against mosquitoes. *J. Invertebr. Pathol.* **1983**, *41*, 143–150. [CrossRef]
46. Meyling, N.V.; Eilenberg, J. Ecology of the entomopathogenic fungi *Beauveria bassiana* and *Metarhizium anisopliae* in temperate agroecosystems: Potential for conservation biological control. *Biol. Control* **2007**, *43*, 145–155. [CrossRef]
47. Pilz, C.; Enkerli, J.; Wegensteiner, R.; Keller, S. Establishment and persistence of the entomopathogenic fungus *Metarhizium anisopliae* in maize fields. *J. Appl. Entomol.* **2011**, *135*, 393–403. [CrossRef]
48. Vidal, S.; Georg-August-University Göttingen, Göttingen, Lower Saxony, Germany. Personal communication. 2015.
49. Strasser, H.; Rauch, H.; Zelger, R. Biological control of adult *Diabrotica* spray experiments with *Metarhizium brunneum* strain BIPESCO 5 under real farm conditions. In Microbial and Nematode Control of Invertebrate Pests, Proceedings of the 16th Meeting IOBC, Tiblisi, Georgia, 11–15 June 2017; Tarasco, E., Jehle, J.A., Burjanadze, M., Ruiu, L., Puza, V., Quesada-Moraga, E., Lopes-Ferber, M., Stephan, D., Eds.; IOBC: Darmstadt, Germany, 2017; Volume 129, pp. 84–87.

Article

# Micro-Landscape Dependent Changes in Arbuscular Mycorrhizal Fungal Community Structure

**Stavros D Veresoglou [1,*,†], Leonie Grünfeld [1,2,†] and Magkdi Mola [1]**

[1] Institut für Biologie, Freie Universität Berlin, Altensteinstr. 6, D-14195 Berlin, Germany; lgreiner@zedat.fu-berlin.de (L.G.); mola.magdi@gmail.com (M.M.)
[2] Berlin-Brandenburg Institute of Advanced Biodiversity Research, D-14195 Berlin, Germany
* Correspondence: sveresoglou@zedat.fu-berlin.de
† These two authors contributed equally.

**Abstract:** The roots of most plants host diverse assemblages of arbuscular mycorrhizal fungi (AMF), which benefit the plant hosts in diverse ways. Even though we understand that such AMF assemblages are non-random, we do not fully appreciate whether and how environmental settings can make them more or less predictable in time and space. Here we present results from three controlled experiments, where we manipulated two environmental parameters, habitat connectance and habitat quality, to address the degree to which plant roots in archipelagos of high connectance and invariable habitats are colonized with (i) less diverse and (ii) easier to predict AMF assemblages. We observed no differences in diversity across our manipulations. We show, however, that mixing habitats and varying connectance render AMF assemblages less predictable, which we could only detect within and not between our experimental units. We also demonstrate that none of our manipulations favoured any specific AMF taxa. We present here evidence that the community structure of AMF is less responsive to spatio-temporal manipulations than root colonization rates which is a facet of the symbiosis which we currently poorly understand.

**Keywords:** Glomeromycota; microbial meta-communities; mycorrhizal mutualistic interactions; null model analyses; plant-soil interactions

**Citation:** Veresoglou, S.D; Grünfeld, L.; Mola, M. Micro-Landscape Dependent Changes in Arbuscular Mycorrhizal Fungal Community Structure. *Appl. Sci.* **2021**, *11*, 5297. https://doi.org/10.3390/app11115297

Academic Editor: Maraike Probst

Received: 20 April 2021
Accepted: 1 June 2021
Published: 7 June 2021

**Publisher's Note:** MDPI stays neutral with regard to jurisdictional claims in published maps and institutional affiliations.



## 1. Introduction

Arbuscular mycorrhizal (AM) associations form direct nutritional symbioses between the roots of most terrestrial plants and a monophyletic group of soil-borne fungi belonging to the phylum Glomeromycota [1]. AM associations have attracted a lot of attention because they can promote net primary productivity (NPP) and agricultural production [2,3]. NPP gains can partially determine how AM fungal communities in plant roots are structured [4–6]. As a result, a lot of the literature addresses practices that likely select for more beneficial communities of Glomeromycota in plant roots (e.g., [7,8]) and environmental parameters and practices that determine AM fungal community structure (e.g., [7,9,10]). An alternative way to ask this question is via questioning how AM fungal diversity varies in space and time (i.e., which entails addressing the fraction of variance which is often classified in models as "unexplained"; [11,12]).

Our general understanding so far is that AM fungal assemblages in the roots are non-random. This has been shown both in relation to null-model analyses [13,14], which assess the degree to which chance exclusively could have generated the observed community table (i.e., the occurrences of AM fungal species across root samples) of the study, and models exploring species-abundance distributions [15,16], which essentially test whether particular groups of species have been more abundant than expected by chance. Many studies observing preferential establishment of AMF taxa in specific habitats also hint towards this direction (e.g., [9,17]). Specific biotic and abiotic parameters of the habitat (besides exerting selectivity to specific AM fungal taxa), however, might also alter our ability to predict (i.e., modify the predictability of) mycorrhizal community structure in nature, but this point remains underexplored. Two syntheses which addressed this question found that anthropogenic disturbances, environmental heterogeneity and a plant host identity (i.e., being a monocotyledon) render AM fungal communities less predictable (i.e., more divergent) than they would have been expected to be by chance alone [18,19]. More recently, Deveautour et al. [20] assayed AM fungal communities in the field to determine the degree to which AM fungal communities diverge with spatial distance but also when sampling from the root systems of the same or from a different plant-host individual. Deveautour et al. [20] observed small differences in AMF community turnover between adjacent neighbouring plants (as compared to sampling from the same individual) but also that AMF community turnover increased for plant individuals further away from each other.

A particular feature of AM fungi is that they are obligate symbionts, meaning that they cannot fulfil their life cycle in the absence of a suitable host. This limits their ability to colonize soil in some environments because their vegetative growth ceases at distances of about 50 cm from the closest colonized root [21]. There is a large body of literature addressing how dispersal constraints modify the community structure of organisms addressing variable types of landscape which can also occur at a micro level such as in soil in which case we can refer to them as micro-landscapes or meta-communities. There is a consensus that meta-communities simultaneously reduce local ($\alpha$-) diversity and increase global ($\gamma$-) diversity because they make local community structure less predictable (e.g., [22]) which potentially allows persistence of less competitive species [23]. This point remains underexplored in relation to AM-associations [24]. Here we present a synthesis from three controlled studies with an overall aim to address how spatial structure in plant mesocosms alters predictability in AM fungal communities. Based on the points we made (e.g., [13,15]), we expected that in all experiments AM fungal communities were non-random (*Hypothesis One*) and that we would observe the highest $\gamma$-diversity in those cases in which the connectance of the patches in the archipelago is lowest (*Hypothesis Two*). Finally, we expected that lowering the connectance of plant and fungal mycorrhizal communities would increase segregation (i.e., the community table becomes more evenly dispersed via weakening pairwise interactions in agreement with the results from Hein et al. [25] showing that strong pairwise interactions promote species aggregation) in Glomeromycota (*Hypothesis Three*). To the best of our understanding, the point that segregation in AM fungal communities could depend on the structure of the micro-landscape has never been addressed in the past for any fungal group and showcases the high potential (because they have an obligate symbiotic lifestyle and are ubiquitous in nature) of using Glomeromycota as model systems in fungal ecology.

## 2. Materials and Methods

### 2.1. Rationale of the Experiments

We worked with large mesocosms (i.e., 90 × 90 × 20 cm) as experimental units to which, for consistency with the meta-community literature, we refer to as archipelagos (Figure 1). Within the mesocosms we established patches (i.e., patches in the form of 8 cm diameter × 20 cm height cylindrical inserts containing 30 μm mesh-covered windows to block root growth but allow growth of fungal hyphae) of vegetated habitat and manipulated the connectance of the patches either by means of distances across patches of the "meta-community" (Experiment One and Experiment Two) or the fertility of the patches within each mesocosm (Experiment Three). At the same time via manipulating the distances of the patches we altered the spatial availability of nutrients in the mesocosms and likely also that (i.e., the spatial distribution) of AMF propagules which were contained in those inserts (and were thus influenced by their spatial arrangement). We anticipated that the lack of prospective hosts between inserts (i.e., patches), over distances of up to 70 cm, hindered dispersal of AMF and would induce meta-community dynamics in our experimental units. The idea of using meta-community theory to model symbiotic systems has been developed and explained in larger detail by Mihaljevic [26] (but see Veresoglou et al. [24] for some likely limitations of the approach in the particular case of AMF communities).

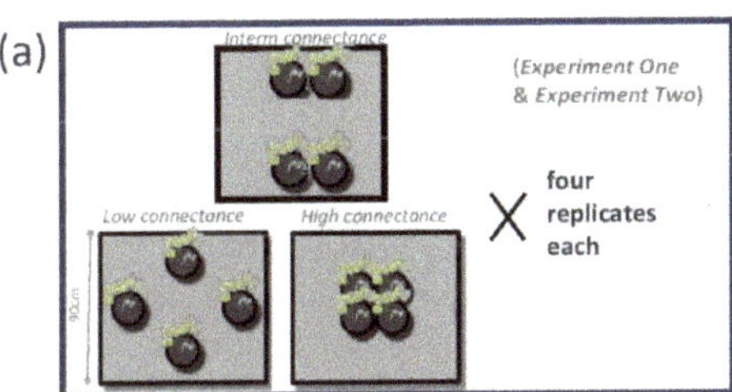


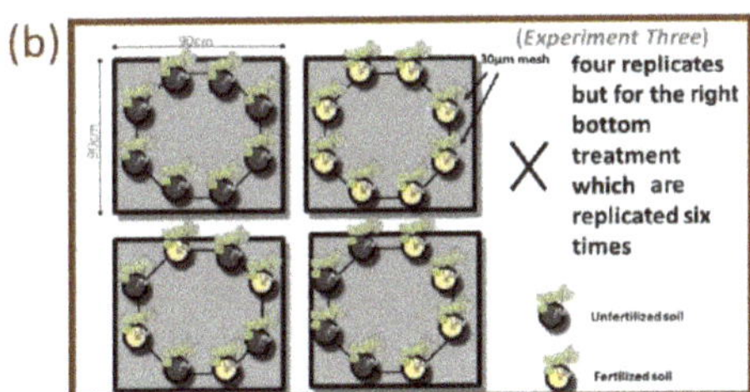


**Figure 1.** Schematic representation of the experimental design of the three experiments. (**a**) In Experiment One and Experiment Two we manipulated the connectance (low; intermediate and high) of four vegetated inserts (dark grey) over an unvegetated soil (sterilized and diluted with sand) matrix (in light grey). In Experiment One we used *Plantago lanceolata* as a host whereas in Experiment Two *Medicago lupulina*. (**b**) In Experiment Three we manipulated the diversity (i.e., only one habitat type; either fertilized or unfertilized or both habitat types) and spatial structure (overdispersed vs. aggregated in the bottom two subpanels) of the vegetated inserts which we describe earlier (Top and bottom left archipelagos/treatments: 4 replicates/were each replicated four times; bottom right archipelagos/treatment: six replicates/ was replicated six times).We used *Medicago lupulina* as a host and the matrix soil was (like in the other experiments) sterilized, mixed with sand and was kept unvegetated (light grey).

### 2.2. Experimental Work

The experimental work on Experiment Two and Experiment Three has been described in detail in Grünfeld et al. ([27]; the two experiments are described there as Experiment One and Experiment Two, respectively; Figure 1). In brief, we carried out three controlled experiments with rectangular mesocosms sized 90 × 90 × 20 cm (width × length × height; Figure 1). Experiment One and Experiment Two used identical experimental designs consisting of four inserts per mesocosm positioned at different distances (three different levels each replicated four times generating archipelagos of low, intermediate and high connectance) from each other but were carried out with different hosts (*Plantago lanceolata* and *Medicago lupulina*; Figure 1a). In Experiment Three we experimented with two different habitats (unfertilized soil and soil fertilized with 1.8 g superphosphate per insert) and the spatial structure of mixtures of them (i.e., aggregated vs. overdispersed spatial structure). In Experiment One some of the *P. lanceolata* roots penetrated the 30 µm mesh barriers and explored the unvegetated compartment. In Experiment Two and Experiment Three we observed differences in AMF colonization across the treatments which we presented in detail in Grünfeld et al. [27].

The soil that was used for the three experiments was collected from a location in northwest Berlin (52.51° N, 13.14° E), had a pH of 6.7 and contained on average 1.75% organic C and 1.3 g $kg^{-1}$ N. The freshly collected soil used for the experiments was stored at room temperature for less than two weeks before setting up the experiments. The soil used to fill the patches was unsterilized providing natural microbiota. The soil used to fill the main compartment of the experimental units was mixed 1:1 with sand and steam-sterilized (99 °C for 2 h) in order to destroy AMF propagules. To each of the inserts we added 200–250 seeds (B&T World Seeds, Aigues-Vives, France) of either *P. laneolata* (Experiment One) or *M. lupulina* (Experiments Two and Three) to approximate a plant density of one seedling per square cm (e.g., [28]).

In the three experiments we used a fully randomized design. Because of the size of the mesocosms it was impossible to re-randomize the experimental units over the duration of the experiment. The temperature in the air-conditioned glasshouse was maintained close to 20 °C. In all three experiments, two weeks after germination of the seedlings, we set up an automatic irrigation system so that the plants were watered daily (over the first two weeks of the experiments watering was carried out manually). We further controlled growth conditions with five soil moisture sensors (ECH20 EC-5 soil moisture sensors and an Em50 data logger, METERs) positioned in three experimental units: in each experimental unit one of the sensors was in the unvegetated compartment and one in one of the inserts. Watering was adjusted so that soil moisture ranged between 60 and 75 % of the water holding capacity. We inspected plant growth daily and removed any unwanted seedlings.

All three experiments were harvested 12 weeks after sowing, respectively, and cleaned root samples (50 mL core) were frozen at −20 °C before DNA extraction. Plant biomass was dried at 60 °C for three days and weighted. Root material from each insert was used to assess root colonization [29]. Soil cores (five per experimental unit with more details in Grünfeld et al. [27]) were used to assay extraradical hyphae in soil. These results are described in Grünfeld et al. [27].

### 2.3. Molecular Analyses and Bioinformatics

Roots from each individual insert per experiment were treated as one sample.Root samples were freeze-dried and homogenized with a Retsch Mixer Mill MM 400 and DNA was extracted from 30 mg ground root material per sample with the DNeasy® PowerPlant® Pro Kit (Qiagen). DNA was amplified with a proofreading polymerase (Kapa HiFi; Kapa Biosystems) and the primer pair NS31-AML2 targeting Glomeromycota (Lee et al., 2008). Thermocycling conditions were as follows: Initial denaturation at 95 °C for 2 min, 35 cycles with first 98 °C for 45 s, then 65 °C for 45 s and 72 °C for 45 s and final elongation at 72 °C for 10 min. The PCR master mix for indexing consisted of 1 µL of the purified polymerase chain reaction (PCR) template, 2.4 µL of the primer mix, 0.25 µL polymerase, 0.5 µL dNTPs (10 µM), 5 µL PCR buffer and 15.85 µL nuclease-free water per 25 µL reaction. Amplicons were purified with the NucleoSpin® gel and PCR clean-up kit (Macherey-Nagel, Düren, Germany) and indexed for MiSeq sequencing by means of an additional PCR with the same conditions as described earlier but with only 15 cycles. Amplicons were purified with magnetic beads (GC Biotech, Alphen aan den Rijn, The Netherlands), and were pooled at equimolar quantities. Sequencing was carried out at the Berlin Center for Genomics in Biodiversity Research (BeGenDiv, Berlin, Germany).

Raw sequences were processed with the UPARSE pipeline [30] with USEARCH v 10.0.240 and default settings and were clustered into phylotypes (i.e., Operational Taxonomic Units - OTUs) at a threshold of 97% sequence similarity. Representative OTU sequences were blasted against MaarjAM [31] and non-specific to Glomeromycota (i.e., >97.5 % similarity or >99 % coverage) OTUs were excluded from further analyses. We then rarefied these to 2200 reads which filtered out two samples from further analyses (i.e., analysis was carried out on the remaining 240 samples).

### 2.4. Statistical Analyses

To address Hypothesis One, stating that AM fungal local communities were non-random, we compared C score (i.e., checkerboards) occurrences in our presence-absence community tables with 1000 randomizations in which we maintained the total number of row sums fixed and the column sums proportional to those of the original community table.

This was carried out through the *sim4* algorithm (i.e., which is appropriate for assessing incomplete lists) [32] which we implemented through the R package EcoSimR [33]. We presented the results in the form of z-score standardized effect sizes (SES) which can be interpreted as (1) random community structure in the case of scores with absolute SES values below 1.96; (2) aggregation for negative SES values below −1.96; and (3) segregation for positive values above 1.96.

To address Hypothesis Two, stating that low connectance promoted a high $\gamma$-diversity in Glomeromycotan communities we used a fixed-effects linear models. We assayed how the experimental design (a categorical predictor with three levels: high connectance vs. intermediate vs. low connectance archipelagos; Figure 1) modified $\gamma$-diversity (response variable) in the experimental units. To further gauge the impact of connectance on $\alpha$- and $\gamma$- diversity we calculated those indices (i.e., local to the inserts and global for the entire mesocosm richness estimates, describing essentially the observed in the resulting community table number of OTUs at each of the two hierarchical levels) for individual inserts and modelled them after a repeated-measures analysis of variance (ANOVA) approach in which we used as response variables the diversity indices and the type of meta-community as predictors with additional error terms to model the nesting of inserts within experimental units. To further address the possibility that the treatments induced differences at a community level we implemented redundancy analyses (RDA) with the Hellinger transformed community tables as responses and the treatments as predictors. Additionally, we carried out an indicator species analysis to assess the degree to which phylotypes preferably established in some spatial designs.

To address Hypothesis Three, stating that low connectance of plant communities increases segregation, we used the Jaccard index (i.e., Jaccard similarity coefficient), defined for any pairwise combination of habitats as the ratio of common species over total number of species, as a metric of similarity across communities. We calculated Jaccard similarities for all pairwise combinations of inserts within individual experimental units. To avoid inflating the degrees of freedom we averaged the similarity coefficients describing the similarity of any given insert across habitats of any particular class (i.e., short distance/long distance/(un)fertilized soil patches). To model similarity coefficients we used a repeated-measures ANOVA approach with the Jaccard coefficients as response variable and a structure identical to the models we used to model $\alpha$- and $\gamma$- diversity.

## 3. Results

### 3.1. Overall Statistics

Alpha diversity varied in the experiments between 6 and 44 phylotypes (i.e., 12–40 in Experiment One; 17 to 44 in Experiment Two and 6 to 41 in Experiment Three; Figure 2). Gamma diversity varied between 30 and 53 phylotypes (i.e., 30–48 in Experiment One; 39 to 52 in Experiment Two; 35 to 53 in Experiment Three). In none of the three experiments could we explain alpha (F values varied between 0.38 and 2.1 with respective $p$ values larger than 0.11) or gamma diversity (F values varied between 0.2 and 1.3 with respective $p$ values larger than 0.3) based on the experimental treatments.

Community differences across the treatments were not significant in any of the three experiment specific RDAs (F values varied between 0.96 and 1.22 with adjective $R^2$ values were in all cases below 0.005). Indicator species analysis yielded inconsistent results and a low occurrence frequency of indicators: there were no indicator OTU in Experiment One, there was a single indicator OTU in Experiment Two specific to low connectance archipelagos (with $p = 0.03$) and there were two OTUs specific to the unfertilized control and one to the fertilized control but not to any of the mixes of them in Experiment Three. Such low frequencies of indicators could have been explained, at least in the case of the two first experiments, by chance.

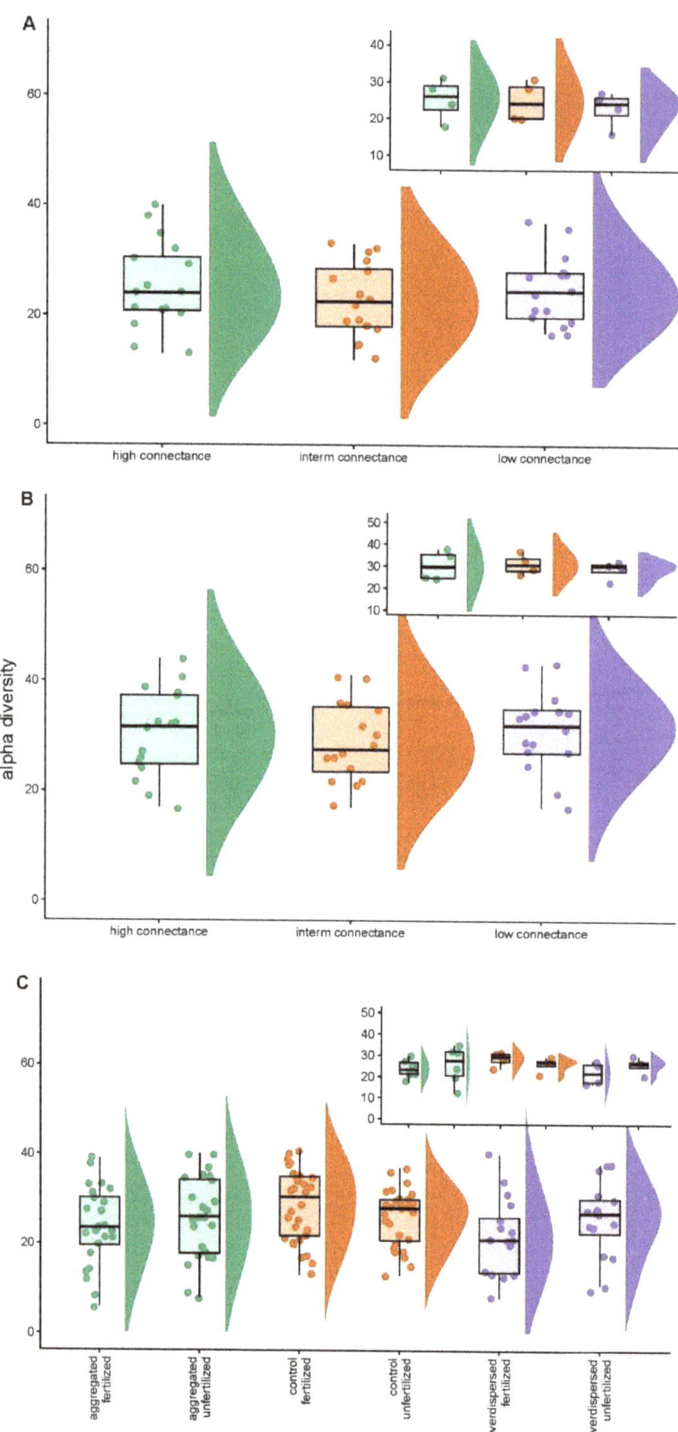


**Figure 2.** Alpha diversity statistics across treatments and habitats in (**A**) Experiment One;

(**B**) Experiment Two; (**C**) Experiment Three. Each experimental unit contained several inserts and we assayed the mycorrhizal community independently for each insert. We observed no differences in alpha diversity in all three experiments. Main panels depict alpha diversity across individual samples whereas the panel inserts show the results after averaging the four (Experiment One and Experiment Two) or eight (Experiment Three) estimates of alpha diversity per experimental unit. Note the lack of differences in relation to alpha diversity. We observed comparable trends for gamma diversity.

### 3.2. Null Model Analyses

All standardized effect size statistics differed from zero and ranged between −9.6 and −20.6 (Experiment One: −9.6; Experiment Two: −10.99 and Experiment Three: −20.66), suggesting community aggregation.

### 3.3. Comparative Analysis of Jaccard Similarities across Experiments

Jaccard similarities did not differ across treatments but within experimental units between short-distance and long-distance inserts in the intermediate connectance treatment of Experiment One ($F_{1,49}$ = 6.3, $p$ = 0.015; Figure 3a; Test 1.1 in the supplementary matterials). There was a comparable trend with Jaccard similarities ($F_{1,49}$= 2.12, $p$ = 0.15) in Experiment Two (Figure 3b; Test 2.1 in the supplementary materials). In Experiment Three, there were differences in Jaccard similarities only between observations within experimental units which differed in their habitat type (i.e., unfertilized vs. fertilized; unfertilized vs. unfertilized; fertilized vs. fertilized; $F_{2,207}$= 4.0, $p$ = 0.02; Figure 3c; Test 3.1 in the supplementary materials). Jaccard similarities were on average larger in the overdispersed treatment compared to the aggregated treatment in Experiment Three (t = −2.03, $p$ = 0.044; Test 3.2 in the supplementary materials).

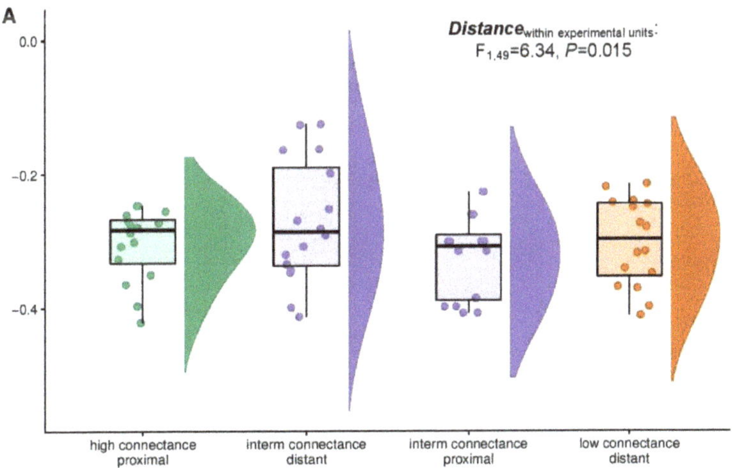


**Figure 3.** *Cont.*

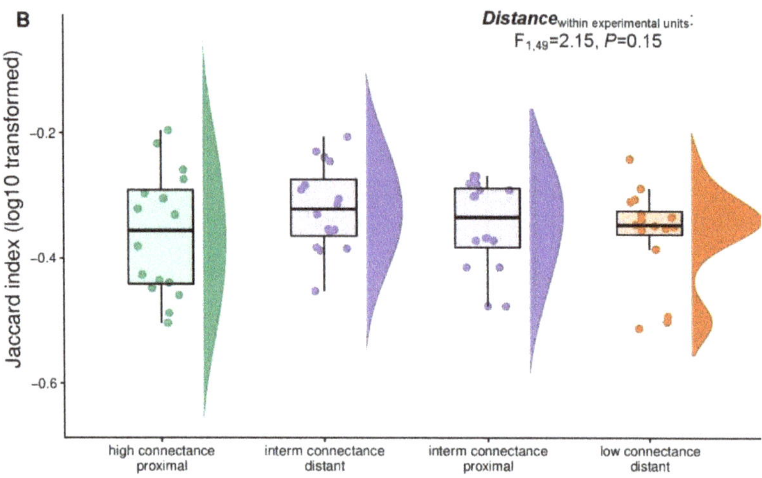


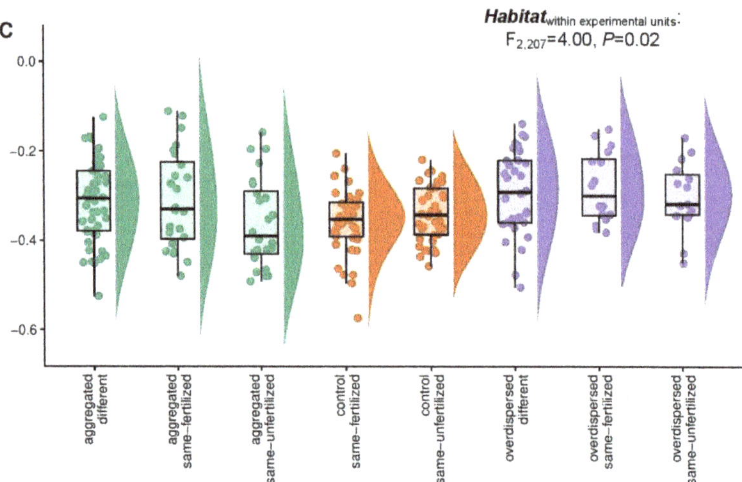


**Figure 3.** Jaccard index statistics from pairwise comparisons of samples within experimental units in (**A**) Experiment One; (**B**) Experiment Two; (**C**) Experiment Three. Note in panels (**A**,**B**) that in intermediate connected spatial arrangements we observe higher Jaccard distances between long-distance compared to short-distance "patches" (in the case of Experiment Two a trend) within-subjects distance effect; $F_{1,49} = 6.34$, $p = 0.015$ in Experiment One; $F_{1,49} = 2.12$, $p = 0.15$ in Experiment Two) and in panel (**C**) that Jaccard distances differed ($F_{2,\,207} = 4$, $p = 0.02$) within experimental units in the overdispersed and aggregated treatment.

We thus observed that within experimental units there were differences in AMF community turnover (assayed with the Jaccard index) which peaked for pairs of distantly placed patches (Experiment One and Experiment Two; as compared to closely placed patches) and pairs of patches containing different habitats (Experiment Three). However, we observed no comparable differences between treatments only containing distantly placed vs. only closely placed patches or high fertility vs. low fertility patches (and this is why the predictor treatment was not significant). A high community turnover, in the absence of diversity differences, is evidence of a lower predictability.

## 4. Discussion

We present evidence from three controlled experiments supporting that small scale micro-landscape heterogeneity (i.e., here describing either experimental units with a low patch-connectance or experimental units containing habitats of different quality) hardly alters diversity patterns in AMF communities. AMF community structure, however, remains non-random. At the same time, we observed that archipelagos combining habitats of both low- and high-connectance (which implies that AMF might have needed to combine traits of long-distance and short-distance dispersal), as well as overdispersed micro-landscapes displayed differences in community turnover (and thus predictability; Test 3.2 in the supplement) across their patches, with pairs of highly connected patches and patches sharing comparable habitats being the most similar to each other. Some conceptual models predict that AMF communities become less random at small (local) spatial scales (i.e., manifested in the form of a low community turnover [34]). With this study we provide experimental evidence that even at such small spatial scales, micro-landscape variability continues to structure AMF communities and can alter their stochasticity (i.e., used here as an opposite to predict).

Our Hypothesis Two stated that we would observe the highest $\gamma$-diversity in the cases in which they were most fragmented, but we observed that the differences across treatments in our experiments were unrelated to AMF diversity. Evidence suggests that AMF richness (either in the form of alpha, here defined as OTUs observed per insert, or gamma diversity, here describing the number of OTUs per mesocosm) stays relatively constant across a range of environmental gradients in AMF systems (e.g., [35–37]) albeit this is not the case with nutrient availability gradients as has been for example shown in Camenzind et al. [38]. It has actually been proposed that plants impose a strong filter on the number of partners they simultaneously associate with [6,39,40], which could determine AMF richness in plant roots. In our experimental set ups, manipulations of the spatial design altered AMF root colonization [27]. The exact reasons why in mixed micro-landscapes we observed a higher root colonization and variable AMF community turnover (which was masked when comparing across less diverse micro-landscapes) are not clear. We suspect that the underlying mechanism relates to alternative growth strategies across AMF taxa. AMF have been proposed to contain two types of hyphae, absorptive and explorative, which differ in their functions (e.g., [41]). Mixed micro-landscapes might necessitate both types of hyphae to be present at high densities which likely weakens interspecific pairwise co-occurrence interactions across AMF species (i.e., pairs of species found together more frequently than expected by chance and pairs of species co-occurring less frequently than expected by chance; e.g., [42]). High densities of both types of hyphae should theoretically result in a higher diversity of pairwise interactions, including many combinations of short-distance dispersers and long-distance dispersers. Also, mixed micro-landscapes could render the benefits that plants acquire from the different AMF species more variable with long-distance dispersers being favoured in some parts of the micro-landscape whereas short-distance ones in others and thus generate conditions with unclear investment optima. In doing so, mixed microlandscapes favour a more diverse set of AMF [6]. An alternative explanation is that within experimental units we could better control for idiosyncratic parameters that can sometimes determine AMF community structure in the early stages, such as the quality and quantity of the AMF propagules and soil moisture settings throughout the experiment. We think that through controlling those idiosyncratic parameters in our within experimental units comparisons, we might had a higher statistical power to detect differences in community turnover (and thus predictability) than across experimental units.

We found support for Hypothesis One that AM fungal communities were non-random which, however, was not surprising. A large body of the mycorrhizal literature supports the idea as we reviewed in the introduction (e.g., [13,14]). What makes our study novel is that across three controlled experiments we found consistent results on a parameter that determines how random AM fungal communities might be micro-landscape structure. We observed differences in community turnover in mixed micro-landscapes (that were masked in their homogenized counterparts) which was higher for distant patches and patches differing in their habitat quality. This observation aligns well with expectations based on meta-community theory [22]. There have only been a few studies so far quantitatively (i.e., assessing effect sizes on the degree of predictability, rather than simply obtaining a qualitative result such as whether the community is segregated) assessing how predictable synthetic microbial systems can be. A recent meta-analysis on the topic examining 21 datasets showed that organic additions make microbial communities less predictable (i.e., more stochastic; [43]) which was later further supported by an additional study [44]. In another study, Fodelianakis et al. [45] showed that evolutionary drift in synthetic bacterial communities rendered them less predictable than in their original cultures. We used here an important for the functioning of ecosystems, system, arbuscular mycorrhizae, to show that also spatial structure can induce less predictable microbial communities and that this happens when we mix different micro-landscape features.

Arbuscular mycorrhizal fungi are most likely to experience dispersal constraints in urban and agricultural landscapes as well as woody habitats [27]. In the case of agricultural landscapes, the growth settings most likely select for short-distance dispersal traits (i.e., there are uniform distances across crop individuals, which ease the proliferation of AMF species from close by patches of AMF diversity). In contrast, in woody habitats the growth settings most likely select for a combination of long- and short-distance dispersal (i.e., distances between AMF-associating plants most likely vary in time and space). Based on the results of our study, plant hosts in woody habitats could, therefore, experience a higher stochasticity in relation to harbouring AMF community structure than other hosts. This might actually benefit AMF-associating plants in forests, in the longer term. Woody plants, in particular, experience a high mortality at early life-stages. If plant fitness to a certain degree depends on the benefits they acquire from associating with AMF (as we suggest in Veresoglou et al. [46] and Grünfeld et al. [27]), stochasticity in AMF community structure could render plant fitness more variable in both time and space and ensure that the surviving individuals are those that associate with strongly mutualistic AMF (e.g., [47]). Further studying parameters that determine stochasticity in mycorrhizal fungal communities, could be key to explaining why and how plant–soil feedback varies in time and space (e.g., [48,49]).

## 5. Conclusions

In conclusion we present evidence that mixing micro-habitat features, such as distances across hosts and fertility levels, makes AMF communities more stochastic (i.e., less predictable). This observation presents a range of opportunities to increase AMF diversity (via facilitating establishment of less competitive species) and hopefully productivity in silviculture and agriculture. Glomeromycota, clearly, present a special case of fungi because of their obligate symbiotic lifestyle, meaning that it is possible to control their spatial structure through manipulating the location of their plant hosts. A follow up question revolves around assessing the degree to which there are comparable patterns in other systems of fungi and the overall consequences for ecosystem functioning.

**Supplementary Materials:** The following are available online at https://www.mdpi.com/article/10.3390/app11115297/s1, Overview (and detailed statistics) of the six tests that had Jaccard coefficients as response variables.

**Author Contributions:** Conceived the study, did the statistics/bioinformatics and prepared the first draft of the paper: S.D.V.; Carried out the experimental work and contributed to the writing of the manuscript: L.G.; Carried out the molecular work described in the study: L.G., M.M.; All authors have read and agreed to the published version of the manuscript.

**Funding:** The authors acknowledge funding from the Deutsche Forschungsgemeinschaft Project Metacorrhiza (VE 736/2-1) awarded to S.D.V.

**Institutional Review Statement:** The study received no institutional reviews.

**Informed Consent Statement:** The study used no humans or animals.

**Data Availability Statement:** The data will be made available in an online repository upon acceptance of the article.

**Acknowledgments:** We want to thank Matthias Rillig and Maraike Probst for constructive comments. The publication of this article was funded by Freie Universität Berlin.

**Conflicts of Interest:** The authors declare no conflict of interest.

## References


1. Brundrett, M.C.; Tedersoo, L. Evolutionary history of mycorrhizal symbioses and global host plant diversity. *New Phytol.* **2018**, *220*, 1108–1115. [CrossRef] [PubMed]
2. Hoeksema, J.D.; Chaudhary, V.B.; Gehring, C.A.; Johnson, N.C.; Karst, J.; Koide, R.T.; Pringle, A.; Zabinski, C.; Bever, J.D.; Moore, J.C.; et al. A meta-analysis of context-dependency in plant response to inoculation with mycorrhizal fungi. *Ecol. Lett.* **2010**, *13*, 394–407. [CrossRef]
3. Zhang, S.; Lehmann, A.; Zheng, W.; You, Z.; Rillig, M.C. Arbuscular mycorrhizal fungi increase grain yields: A meta-analysis. *New Phytol.* **2019**, *222*, 543–555. [CrossRef]
4. Van Der Heijden, M.G.A.; Klironomos, J.N.; Ursic, M.; Moutoglis, P.; Streitwolf-Engel, R.; Boller, T.; Wiemken, A.; Sanders, I.R. Mycorrhizal fungal diversity determines plant biodiversity, ecosystem variability and productivity. *Nature* **1998**, *396*, 69–72. [CrossRef]
5. Maherali, H.; Klironomos, J.N. Influence of phylogeny on fungal community assembly and ecosystem functioning. *Science* **2007**, *316*, 1746–1748. [CrossRef] [PubMed]
6. Kiers, E.T.; Duhamel, M.; Beesetty, Y.; Mensah, J.A.; Franken, O.; Verbruggen, E.; Fellbaum, C.R.; Kowalchuk, G.A.; Hart, M.M.; Bago, A.; et al. Reciprocal rewards stabilize cooperation in the mycorrhizal symbiosis. *Science* **2011**, *333*, 880–883. [CrossRef]
7. Rillig, M.C.; Sosa-Hernández, M.A.; Roy, J.; Aguilar-Trigueros, C.A.; Vályi, K.; Lehmann, A. Towards an integrated mycorrhizal technology: Harnessing mycorrhiza for sustainable intensification in agriculture. *Front. Plant Sci.* **2016**, *7*, 1625. [CrossRef]
8. Pánková, H.; Lepinay, C.; Rydlová, J.; Voková, A.; Janoušková, M.; Dostálek, T.; Münzbergová, Z. Arbuscular mycorrhizal fungi and associated microbial communities from dry grassland do not improve plant growth on abandoned field soil. *Oecologia* **2018**, *186*, 677–689. [CrossRef]
9. Egerton-Warburton, L.M.; Johnson, N.C.; Allen, E.B. Mycorrhizal community dynamics following nitrogen fertilization: A cross-site test in five grasslands. *Ecol. Monogr.* **2007**, *77*, 527–544. [CrossRef]
10. Veresoglou, S.D.; Caruso, T.; Rillig, M.C. Modelling the environmental and soil factors that shape the niches of two common arbuscular mycorrhizal fungal families. *Plant Soil* **2013**, *368*, 507–518. [CrossRef]
11. Dumbrell, A.J.; Ashton, P.D.; Aziz, N.; Feng, G.; Nelson, M.; Dytham, C.; Fitter, A.H.; Helgason, T. Distinct seasonal assemblages of arbuscular mycorrhizal fungi revealed by massively parallel pyrosequencing. *New Phytol.* **2011**, *190*, 794–804. [CrossRef] [PubMed]
12. Davison, J.; Öpik, M.; Zobel, M.; Vasar, M.; Metsis, M.; Moora, M. Communities of arbuscular mycorrhizal fungi detected in forest soil are spatially heterogeneous but do not vary throughout the growing season. *PLoS ONE* **2012**, *7*, e41938. [CrossRef]
13. Davison, J.; Öpik, M.; Daniell, T.J.; Moora, M.; Zobel, M. Arbuscular mycorrhizal fungal communities in plant roots are not random assemblages. *FEMS Microbiol. Ecol.* **2011**, *78*, 103–115. [CrossRef] [PubMed]
14. Horn, S.; Hempel, S.; Verbruggen, E.; Rillig, M.C.; Caruso, T. Linking the community structure of arbuscular mycorrhizal fungi and plants: A story of interdependence? *ISME J.* **2017**, *11*, 1400–1411. [CrossRef]
15. Dumbrell, A.J.; Nelson, M.; Helgason, T.; Dytham, C.; Fitter, A.H. Idiosyncrasy and overdominance in the structure of natural communities of arbuscular mycorrhizal fungi: Is there a role for stochastic processes? *J. Ecol.* **2010**, *98*, 419–428. [CrossRef]
16. Unterseher, M.; Jumpponen, A.; Öpik, M.; Tedersoo, L.; Moora, M.; Dormann, C.F.; Schnittler, M. Species abundance distributions and richness estimations in fungal metagenomics—Lessons learned from community ecology. *Mol. Ecol.* **2011**, *20*, 275–285. [CrossRef]
17. Chen, Y.L.; Xu, Z.W.; Xu, T.L.; Veresoglou, S.D.; Yang, G.W.; Chen, B.D. Nitrogen deposition and precipitation induced phylogenetic clustering of arbuscular mycorrhizal fungal communities. *Soil Biol. Biochem.* **2017**, *115*, 233–242. [CrossRef]
18. Caruso, T.; Powell, J.R.; Rillig, M.C. Compositional divergence and convergence in local communities and spatially structured landscapes. *PLoS ONE* **2012**, *7*, 1115–1124. [CrossRef] [PubMed]
19. Powell, J.R.; Bennett, A.E. Unpredictable assembly of arbuscular mycorrhizal fungal communities. *Pedobiologia* **2016**, *59*, 11–15. [CrossRef]
20. Deveautour, C.; Donn, S.; Bennett, A.E.; Power, S.; Powell, J.R. Variability of arbuscular mycorrhizal fungal communities within the root systems of individual plants is high and influenced by host species and root phosphorus. *Pedobiologia* **2021**, *84*, 150691. [CrossRef]
21. Klironomos, J.N.; Moutoglis, P. Colonization of nonmycorrhizal plants by mycorrhizal neighbours as influenced by the collembolan, Folsomia candida. *Biol. Fertil. Soils* **1999**, *29*, 277–281. [CrossRef]

22. Hubbell, S.P. *The Unified Neutral Theory of Biodiversity and Biogeography*; Monographs of Population Biology; Princeton University Press: Princeton, NJ, USA, 2001; Volume 32.
23. Cadotte, M.W. Competition-colonization trade-offs and disturbance effects at multiple scales. *Ecology* **2007**, *88*, 823–829. [CrossRef]
24. Veresoglou, S.D.; Caruso, T.; Rillig, M.C. Metacommunities and symbiosis: Hosts of challenges. *Trends Ecol. Evol.* **2012**, *27*, 588–589. [CrossRef]
25. Hein, C.L.; Öhlund, G.; Englund, G. Fish introductions reveal the temperature dependence of species interactions. *Proc. R. Soc. B Biol. Sci.* **2013**, *281*. [CrossRef]
26. Mihaljevic, J.R. Linking metacommunity theory and symbiont evolutionary ecology. *Trends Ecol. Evol.* **2012**, *27*, 323–329. [CrossRef]
27. Grünfeld, L.; Wulf, M.; Rillig, M.C.; Manntschke, A.; Veresoglou, S.D. Neighbours of arbuscular-mycorrhiza associating trees are colonized more extensively by arbuscular mycorrhizal fungi than their conspecifics in ectomycorrhiza dominated stands. *New Phytol.* **2019**, *227*, 10–13. [CrossRef] [PubMed]
28. Scotton, M. Mountain grassland restoration: Effects of sowing rate, climate and soil on plant density and cover. *Sci. Total Environ.* **2019**, *651*, 3090–3098. [CrossRef]
29. McGonigle, T.P.; Miller, M.H.; Evans, D.G.; Fairchild, G.L.; Swan, J.A. A new method which gives an objective measure of colonization of roots by vesicular- Arbuscular mycorrhizal fungi. *New Phytol.* **1990**, *115*, 495–501. [CrossRef]
30. Edgar, R.C. UPARSE: Highly accurate OTU sequences from microbial amplicon reads. *Nat. Methods* **2013**, *10*, 996–998. [CrossRef]
31. Öpik, M.; Vanatoa, A.; Vanatoa, E.; Moora, M.; Davison, J.; Kalwij, J.M.; Reier, Ü.; Zobel, M. The online database MaarjAM reveals global and ecosystemic distribution patterns in arbuscular mycorrhizal fungi (Glomeromycota). *New Phytol.* **2010**, *188*, 223–241. [CrossRef] [PubMed]
32. Gotelli, N.J. Null model analysis of species co-occurrence patterns. *Ecology* **2000**, *81*, 2606–2621. [CrossRef]
33. Gotelli, N.; Hart, E.; Ellison, A.; Hart, M.E. Package 'EcoSimR'—Null Model Analysis for Ecological Data. *R Packag.* 2015.
34. Vályi, K.; Mardhiah, U.; Rillig, M.C.; Hempel, S. Community assembly and coexistence in communities of arbuscular mycorrhizal fungi. *ISME J.* **2016**, *10*, 2341–2351. [CrossRef] [PubMed]
35. Lekberg, Y.; Schnoor, T.; Kjøller, R.; Gibbons, S.M.; Hansen, L.H.; Al-Soud, W.A.; Sørensen, S.J.; Rosendahl, S. 454-Sequencing Reveals Stochastic Local Reassembly and High Disturbance Tolerance Within Arbuscular Mycorrhizal Fungal Communities. *J. Ecol.* **2012**, *100*, 151–160. [CrossRef]
36. Kivlin, S.N.; Hawkes, C.V. Tree species, spatial heterogeneity, and seasonality drive soil fungal abundance, richness, and composition in Neotropical rainforests. *Environ. Microbiol.* **2016**, *18*, 4662–4673. [CrossRef] [PubMed]
37. Maitra, P.; Zheng, Y.; Chen, L.; Wang, Y.L.; Ji, N.N.; Lü, P.P.; Gan, H.Y.; Li, X.C.; Sun, X.; Zhou, X.H.; et al. Effect of drought and season on arbuscular mycorrhizal fungi in a subtropical secondary forest. *Fungal Ecol.* **2019**, *41*, 107–115. [CrossRef]
38. Camenzind, T.; Hempel, S.; Homeier, J.; Horn, S.; Velescu, A.; Wilcke, W.; Rillig, M.C. Nitrogen and phosphorus additions impact arbuscular mycorrhizal abundance and molecular diversity in a tropical montane forest. *Glob. Chang. Biol.* **2014**, *20*, 3646–3659. [CrossRef] [PubMed]
39. Veresoglou, S.D.; Halley, J.M. A model that explains diversity patterns of arbuscular mycorrhizas. *Ecol. Model.* **2012**, *231*, 146–152. [CrossRef]
40. Hammer, E.C.; Pallon, J.; Wallander, H.; Olsson, P.A. Tit for tat? A mycorrhizal fungus accumulates phosphorus under low plant carbon availability. *FEMS Microbiol. Ecol.* **2011**, *76*, 236–244. [CrossRef]
41. Staddon, P.L.; Ramsey, C.B.; Ostle, N.; Ineson, P.; Fitter, A.H. Rapid turnover of hyphae of mycorrhizal fungi determined by AMS microanalysis of 14C. *Science* **2003**, *300*, 1138–1140. [CrossRef]
42. Bar-Massada, A. Complex relationships between species niches and environmental heterogeneity affect species co-occurrence patterns in modelled and real communities. *Proc. R. Soc. B Biol. Sci.* **2015**, *282*, 20150927. [CrossRef]
43. Ning, D.; Deng, Y.; Tiedje, J.M.; Zhou, J. A general framework for quantitatively assessing ecological stochasticity. *Proc. Natl. Acad. Sci. USA* **2019**, *116*, 16892–16898. [CrossRef]
44. Silva, M.O.D.; Pernthaler, J. Biomass addition alters community assembly in ultrafiltration membrane biofilms. *Sci. Rep.* **2020**, *10*, 1–10. [CrossRef] [PubMed]
45. Fodelianakis, S.; Valenzuela-Cuevas, A.; Barozzi, A.; Daffonchio, D. Direct quantification of ecological drift at the population level in synthetic bacterial communities. *ISME J.* **2021**, *15*, 55–66. [CrossRef] [PubMed]
46. Veresoglou, S.D.; Wulf, M.; Rillig, M.C. Facilitation between woody and herbaceous plants that associate with arbuscular mycorrhizal fungi in temperate European forests. *Ecol. Evol.* **2017**, *7*, 1181–1189. [CrossRef] [PubMed]
47. Johnson, D.W.; Grorud-Colvert, K.; Sponaugle, S.; Semmens, B.X. Phenotypic variation and selective mortality as major drivers of recruitment variability in fishes. *Ecol. Lett.* **2014**, *17*, 743–755. [CrossRef]
48. Kadowaki, K.; Yamamoto, S.; Sato, H.; Tanabe, A.S.; Hidaka, A.; Toju, H. Mycorrhizal fungi mediate the direction and strength of plant–soil feedbacks differently between arbuscular mycorrhizal and ectomycorrhizal communities. *Commun. Biol.* **2018**, *1*, 196. [CrossRef]
49. Liang, M.; Johnson, D.; Bursem, F.R.P.; Yu, S.; Fang, M.; Taylor, J.D.; Taylor, A.F.S.; Helgason, T.; Liu, X. Soil fungal networks maintain local dominance of ectomycorrhizal trees. *Nat. Commun.* **2020**, *11*, 2636. [CrossRef] [PubMed]

Article

# Can We Use Functional Annotation of Prokaryotic Taxa (FAPROTAX) to Assign the Ecological Functions of Soil Bacteria?

**Chakriya Sansupa [1,2,†], Sara Fareed Mohamed Wahdan [2,3,†], Shakhawat Hossen [2,4,†], Terd Disayathanoowat [1,5,6,*], Tesfaye Wubet [7,8,‡] and Witoon Purahong [2,*,‡]**

1 Department of Biology, Faculty of Science, Chiang Mai University, Chiang Mai 50200, Thailand; chakriya_s@cmu.ac.th
2 Department of Soil Ecology, UFZ-Helmholtz Centre for Environmental Research, 06120 Halle (Saale), Germany; sara-fareed-mohamed.wahdan@ufz.de (S.F.M.W.); shakhawat.hossen@ufz.de (S.H.)
3 Department of Botany, Faculty of Science, Suez Canal University, Ismailia 41522, Egypt
4 Institute of Ecology and Evolution, Friedrich-Schiller-Universität Jena, 07743 Jena, Germany
5 Research Center in Bioresources for Agriculture, Industry and Medicine, Chiang Mai University, Chiang Mai 50200, Thailand
6 Research Center of Microbial Diversity and Sustainable Utilization, Chiang Mai University, Chiang Mai 50200, Thailand
7 Department of Community Ecology, UFZ-Helmholtz Centre for Environmental Research, 06120 Halle (Saale), Germany; tesfaye.wubet@ufz.de
8 German Centre for Integrative Biodiversity Research (iDiv), Halle-Jena-Leipzig, 04103 Leipzig, Germany
* Correspondence: terd.dis@cmu.ac.th (T.D.); witoon.purahong@ufz.de (W.P.)
† These authors contributed equally to this work.
‡ These authors are joint senior authors.

**Abstract:** FAPROTAX is a promising tool for predicting ecological relevant functions of bacterial and archaeal taxa derived from 16S rRNA amplicon sequencing. The database was initially developed to predict the function of marine species using standard microbiological references. This study, however, has attempted to access the application of FAPROTAX in soil environments. We hypothesized that FAPROTAX was compatible with terrestrial ecosystems. The potential use of FAPROTAX to assign ecological functions of soil bacteria was investigated using meta-analysis and our newly designed experiments. Soil samples from two major terrestrial ecosystems, including agricultural land and forest, were collected. Bacterial taxonomy was analyzed using Illumina sequencing of the 16S rRNA gene and ecological functions of the soil bacteria were assigned by FAPROTAX. The presence of all functionally assigned OTUs (Operation Taxonomic Units) in soil were manually checked using peer-reviewed articles as well as standard microbiology books. Overall, we showed that sample source was not a predominant factor that limited the application of FAPROTAX, but poor taxonomic identification was. The proportion of assigned taxa between aquatic and non-aquatic ecosystems was not significantly different ($p > 0.05$). There were strong and significant correlations ($\sigma = 0.90$–$0.95$, $p < 0.01$) between the number of OTUs assigned to genus or order level and the number of functionally assigned OTUs. After manual verification, we found that more than 97% of the FAPROTAX assigned OTUs have previously been detected and potentially performed functions in agricultural and forest soils. We further provided information regarding taxa capable of N-fixation, P and K solubilization, which are three main important elements in soil systems and can be integrated with FAPROTAX to increase the proportion of functionally assigned OTUs. Consequently, we concluded that FAPROTAX can be used for a fast-functional screening or grouping of 16S derived bacterial data from terrestrial ecosystems and its performance could be enhanced through improving the taxonomic and functional reference databases.

**Keywords:** soil; FAPROTAX; functional annotation; bacterial function

**Citation:** Sansupa, C.; Wahdan, S.F.M.; Hossen, S.; Disayathanoowat, T.; Wubet, T.; Purahong, W. Can We Use Functional Annotation of Prokaryotic Taxa (FAPROTAX) to Assign the Ecological Functions of Soil Bacteria?. *Appl. Sci.* **2021**, *11*, 688. https://doi.org/10.3390/app11020688

Received: 26 November 2020
Accepted: 5 January 2021
Published: 12 January 2021

## 1. Introduction

Microbes are known as engines of an ecosystem as their growth and metabolisms drive various biogeochemical cycles and mediate many ecological processes such as decomposing organic compounds, solubilizing mineral substances and promoting plant performance [1]. Moreover, microbes also play important roles in removing toxic pollutants and chemical contaminants. For instance, microorganisms transform aromatic compounds into harmless metabolites or less/nontoxic forms [2]. Other harmful metabolites that are converted by microorganisms include hydrocarbon compounds [3], heavy metals [4] and other chemicals with excessive concentrations in an environment [5–7].

Various tools have been developed for the prediction of ecological functions of microbial taxa derived from amplicon-based next-generation sequencing data. These tools allow us to investigate both community and functional composition of microbes. The usefulness of these tools depends on thorough and global data on microbial community and functions, which would provide deeper insight for microbial ecological research and could be a low-cost alternative to metagenomic sequencing [8,9]. As a result, many functional prediction tools were generated for both prokaryotic and eukaryotic microorganisms. For example, FUNGuild is a typical functional prediction tool for fungi, providing guild characteristics of the detected taxa, such as saprotroph, pathogen, decomposer or lichenivorous fungi, based on their taxonomic identity [10]. Other tools such as phylogenetic investigation of communities by reconstruction of unobserved states (PICRUSt) [11,12], pathway prediction by phylogenetic placement (PAPRICA) [13], predicting functional profiles from metagenomic 16S rRNA data (Tax4Fun) [8] and functional annotation of prokaryotic taxa (FAPROTAX) [9] were developed to predict bacterial and archaeal functions. The former three predict the functions based on gene content of detected taxa, whereas the latter is the only tool that uses experimental data on culturable taxa to identify functional groups, metabolic phenotypes or ecologically relevant functions [9]. Furthermore, FAPROTAX may be more preferable for functional prediction of the biogeochemical cycle of environmental samples [9,14,15].

The FAPROTAX is a database that maps bacterial or archaeal taxa to metabolic or ecologically relevant functions (i.e., nitrogen fixation, sulfate respiration or hydrocarbon degradation) using literature on culture representatives. This means that if all cultured members of a taxon can perform a function, the function will be assigned to all members of this taxon (cultured and uncultured). The FAPROTAX provides a python script to convert OTUs tables into functional tables based on taxa identified in a sample and functional phenotype of each taxon in the FAPROTAX database. This database was initially built for a study on marine environments, containing over 80 functions and 4600 taxonomic details of bacteria and archaea from oceans [9]. However, FAPROTAX used standard references (in Bergey's manual of systematic bacteriology [16–20]. The prokaryotes [21] and International Journal of Systematic and Evolutionary Microbiology [22]) for general bacteria and archaea living in both aquatic and terrestrial ecosystems. This implies that the functional assignments are not completely dependent on habitats (aquatic vs. terrestrial systems) and highly dependent on taxonomic information at genus, species or strain levels of a particular bacterial and archaeal taxon. Consequently, many studies have used FAPROTAX for functional annotation of bacteria and archaea on various ecosystems including aquatic [9,23–27] and terrestrial systems [28–31], as well as human and animal microbiome [14,32].

Despite the compatibility to the soil system, FAPROTAX is still not widely utilized for functional annotation of soil bacteria and archaea because there has been no effort to test the capability of FAPROTAX in soil. Some important questions remain, for example, (i) what proportions of bacterial and archaeal taxa are living in soil that are successfully annotated using FAPROTAX (as compared to aquatic systems) and (ii) are functional annotations using FAPROTAX accurate in soil systems?

This study aimed to investigate the potential use of FAPROTAX for bacterial functional annotation in non-aquatic ecosystems, specifically in soil. We used both published and newly prepared soil microbial datasets of three ecosystems (mangrove, agriculture and forest). In total, four datasets, including mangrove rhizosphere soil, agricultural bulk soil, agricultural rhizosphere soil and forest soil, were processed. We (i) tested the differences in functional annotated capacity between non-aquatic and aquatic ecosystem using published articles (meta-analysis), (ii) tested the accuracy of FAPROTAX annotation in soil systems by manually checking both appearance and functional performance in soils of all functionally assigned bacterial and archaeal operational taxonomic units (OTUs) with previously published literature and, (iii) tested additional options to improve the functional assignment capacity of FAPROTAX in soils by using relevant references (case study: nitrogen-fixing bacteria in bulk and rhizosphere soils of *Trifolium pratense*).

## 2. Materials and Methods

### 2.1. Sample Collection

We set up experiments to evaluate the suitability of FAPROTAX to assign functions of soil bacterial and archaeal OTUs. Soil samples from two major terrestrial ecosystems, agricultural grassland and forest were collected. Furthermore, one published dataset on rhizosphere soil of *Rhizophora stylosa* was also added in this study [33]. Mangrove ecosystems are considered as the interface between aquatic and terrestrial ecosystems [34]; thus, the mangrove soil samples were used as the borderline between the aquatic and terrestrial ecosystems.

In the agricultural grassland ecosystem, bulk soil and rhizosphere soil of *Trifolium pratense* were taken from five experimental plots of extensively used meadow (ambient treatment) at the Global Change Experimental Facility (GCEF), Germany (51°22′60″ N, 11°50′60″ E) [35,36]. These experimental plots are managed as described in Schädler et al. [35], where grasses and legumes growing in the plot are used to feed livestock. In this study, three healthy *pratense* plants, which represent three subsamples, were taken from each experimental plot. Bulk soil and rhizosphere soil were collected following the protocol as described by Barillot et al. [37]. Briefly, bulk soil attached to the root was first removed by vigorous shaking, then rhizosphere soil was collected by shaking the root in PCR-grade water. Three subsamples of bulk soil and rhizosphere soil collected from the same plot were pooled into one composite sample for each soil fraction. Overall, five composite samples of bulk soil and five composite samples of rhizosphere soil (representing five true replicates) were obtained.

In the forest ecosystem, soil samples were taken from a bamboo-deciduous forest [38] which was dominated by *Dendrocalamus membranaceous* and *Bambusa bambos*, in northern Thailand (18°32′23″ N, 99°34′47″ E). In detail, five replicated plots (5 × 5 $m^2$) > 20 m apart were selected. Five soil subsamples were collected to 10 cm depth in each plot using an auger with 10 cm in diameter. The subsamples were then pooled into one composite sample. In this study, all composite soil samples were homogenized and sieved (2 mm) to remove stones, roots, macrofauna, and litter.

For the mangrove ecosystem, we used the dataset that has previously been published by Purahong et al. [33]. In detail, this dataset consisted of 5 replicated samples of rhizosphere soil of *R. stylosa* located in wetland at Pingtung County, southern Taiwan (22°26′17.6″ N, 120°29′29.6″ E). Sample collection was described in detail in Purahong et al. [33]. Briefly, five healthy, mature *stylosa* trees were selected, then four subsamples of rhizosphere soil around each selected tree were collected, pooled and sieved to make a composite sample.

All soil samples, including those from agricultural grassland, deciduous forest and mangrove ecosystems, were kept at −20 °C for further analyses.

### 2.2. DNA Extraction, Sequencing and Taxonomy and Functional Assignment

DNA samples of both agricultural soils were extracted by QIAGEN DNeasy PowerSoil kit, whereas those of forest soil were extracted by NucleoSpin® Soil DNA extraction kit. PCR amplification of the bacterial V3-V4 regions was conducted using the bacterial primer pair Bact341F (5′-CCTACGGGNGG-CWGCAG-3′) and Bact785R (5′-GACTACHVGGGTATCTAATCC-3′) [39]. Amplicons were sequenced using an Illumina MiSeq platform and V3 Chemistry (Illumina). All amplification and sequencing steps were performed at RTL Genomics (Lubbock, TX, USA).

Bioinformatics was proceeded on MOTHUR 1.33.3 [40] following *Standard Operating Procedure* (SOP) custom analysis workflow. Raw reads, overlapping more than 20 base pairs, were first assembled to generate paired-end reads, then filtered to get high-quality reads, containing at least 200 bp length and having a minimum average quality of 25 Phred score. Chimeric sequences were checked using UCHIME in de novo mode [41], as implemented in MOTHUR and removed from the datasets. The cleaned sequences were clustered at 97% sequence identity and assigned taxonomy using the SILVA 132 database for bacterial 16S rRNA gene [42]. Ecologically relevant functions were then assigned to all detected OTUs using FAPROTAX [9]. In detail, a taxon (e.g., strain, species or genus) was annotated to certain functions if there was a literature report on a culture representative of the taxon that performed the functions. For example, if all cultured species of a genus were previously reported as sulfate reducers, all detected taxa belonging to the genus were also considered as sulfate reducers. All database and assignment instructions are available at http://www.loucalab.com/archive/FAPROTAX/. Lastly, rare OTUs (singletons to tripletons), which could potentially originate from sequencing error, were removed. The remaining reads were normalized to the minimum read count per sample of each dataset. Final OTU tables of all datasets are available in supplementary Tables S1.1–S1.3. The raw sequences are available in the National Center for Biotechnology Information (NCBI) under BioProject accession number: PRJNA646011 for forest and agricultural soils and PRJNA554586 for mangrove soil.

### 2.3. *Validation of FAPROTAX*

#### 2.3.1. Peer-Reviewed Publications: Difference in Functional Assignment Percentage of Aquatic and Non-Aquatic Samples.

The functional assignment percentage (proportion of number of OTUs assigned by FAPROTAX to total detected OTUs in each study) presented in peer-reviewed publications was gathered and divided based on sampling source into 2 main groups, including aquatic and non-aquatic data (Table S3). Differences in the functional assignment percentage over aquatic and non-aquatic data were tested using the Mann-Whitney U test in PAST program v. 2.17c [43]. Shapiro-Wilk and Fligner-Killeen were used to test normality and equality of variances between the two groups.

#### 2.3.2. Taxonomic Identification and FAPROTAX Assignment

The datasets, including twenty soil samples derived from agricultural-bulk soil ($n$ = 5), agricultural-rhizosphere soil ($n$ = 5), forest soil ($n$ = 5) and mangrove soil ($n$ = 5), were used to test the correlation between number of functionally assigned OTUs and number of OTUs identified to different taxonomic ranks (Genus, Order and Phylum). Firstly, the normality of these data sets was analyzed using Shapiro-Wilk test. We detected non-normal distribution in some data sets ($p < 0.05$); thus, Spearman's rank correlation method was used to test the correlation between number of functionally assigned OTUs and number of OTUs identified to each taxonomic rank. The correlation method was analyzed using stat_cor function in ggpubr package [44]. The simple linear regressions, showing the relationship between the two variables, were plotted with *ggscatter* function of the *ggpubr* package. These correlation analyses were run on R (version 3.6.2) [45].

#### 2.3.3. Validation of FAPROTAX on Soil Bacteria

Bacterial taxa provided in a file called "FAPROTAX.txt", which is a database for FAPROTAX analysis (http://www.loucalab.com/archive/FAPROTAX/lib/php/index.php?section=Download), were randomly selected and habitat of the selected taxa was identified using references cited in the database. This action indicated that prokaryotic taxa and functional results in the FAPROTAX database have been obtained not only from aquatic samples but also from non-aquatic environments (an example is presented in Table S4) which lead to the promising application of FAPROTAX in soil samples.

Subsequently, four datasets, including agricultural-bulk soil, agricultural-rhizosphere soil, forest soil and mangrove soil, were used as examples to validate the suitability of FAPROTAX applied to soil samples. In this section, OTUs that were functionally assigned by FAPROTAX were used (Tables S2.1–S2.3). In detail, FAPROTAX assigned OTUs was first checked on the appearance of each taxon in soil habitat using peer-reviewed publications. If a taxon was previously reported in soil systems, we confirmed the appearance by using the word "Yes". On the other hand, if there was no record of a taxon in soil systems, we used the word "No" (Tables S2.1–S2.3: column named "*Confirmation living in soil*"). Secondly, functional performance in soil systems of the FAPROTAX assigned OTUs was manually checked using a similar procedure as FAPROTAX database (a taxon was assigned to certain functions if there was a literature report on a representative of the taxon that performed the functions). The functional performance was confirmed by using "Yes" when particular OTUs have been reported and performed the function in soil, whereas "No" was used for those with no record of functional performance in soil (Tables S2.1–S2.3: column named "*Confirmation on functions in soil habitat*"). The performance of FAPROTAX assignment in soil sample was indicated by number of FAPROTAX assigned OTUs that were both available and performed the assigned function in soil.

#### 2.3.4. Additional Literature on Soil Functions

Bacterial taxa that were identified as a driver on phosphate solubilization, potassium solubilization and nitrogen fixation were provided (Table 1). These taxa and functions were overlooked from FAPROTAX database (file named "FAPROTAX.txt" available at FAPROTAX webpage). Subsequently, the advantage of the literature added was tested using soil samples from agricultural bulk soil and rhizosphere soil of *T. pratense* datasets. In detail, the nitrogen fixation function was manually assigned to a particular OTU in the datasets when that OTU was identified as one of the taxa in Table 1 (nitrogen fixation). Then, the number of manually assigned OTUs was counted and compared with that of OTUs assigned to nitrogen fixation by FAPROTAX.

**Table 1.** A list of bacterial taxa capable of nitrogen fixation, phosphate and potassium solubilization.

| Functions | Taxa | References |
| --- | --- | --- |
| Nitrogen fixation | *Allorhizobium, Azorhizobium, Bradyrhizobium, Ensifer, Mesorhizobium, Rhizobium, Aminobacter, Devosia, Methylobacterium, Microvirga, Ochrobactrum, Phyllobacterium, Shinella, Burkholderia, Paraburkholderia, Cupriavidus, Bacillus safencis, Bacillus licheniformis, Bacillus cereus, Bacillus megaterium, Bacillus aerophilus, Bacillus flexus, Bacillus oceanisediminis, Bacillus circulans, Bacillus aerophilus, Bacillus subtilis* | [46,47] |
| Potassium solubilization | *Bacillus mucilaginosus, Bacillus circulanscan, Bacillus edaphicus, Bacillus megaterium, Acidithiobacillus ferrooxidans, Enterobacter hormaechei, Paenibacillus mucilaginosus, Paenibacillus glucanolyticus* | [48,49] |

**Table 1.** *Cont.*

| Functions | Taxa | References |
|---|---|---|
| Phosphate solubilization | *Aerobactor aerogenes, Actinomadura oligospora, Azospirillum brasilense, Bacillus circulans, Bacillus cereus, Bacillus fusiformis, Bacillus pumils, Bacillus megaterium, Bacillus mycoides, Bacillus polymyxa, Bacillus coagulans, Bacillus chitinolyticus, Bacillus subtilis, Pseudomonas putida, Pseudomonas striata, Pseudomonas fluorescens, Pseudomonas calcis, Escherichia intermedia, Enterobacter asburiae, Serratia phosphoticum, Thiobacillus ferroxidans, Thiobacillus thioxidans, Rhizobium meliloti, Bacillus pulvifaciens, Bacillus sircalmous, Pseudomonas canescens, Pseudomonas fluorescens, Pantoea agglomerans, Rhizobium meliloti, Rhizobium leguminosarum, Mesorhizobium mediterraneum, Acinetobacter rhizosphaerae, Streptomyces albus, Streptomyces cyaneus, Streptoverticillium album, Azotobacter chroococcum* | [50,51] |

## 3. Results

### 3.1. Assignment Percentage of Aquatic and Non-Aquatic Samples Based on Previous Studies

We found no significant difference between assignment percentage (proportion of FAPROTAX assigned OTUs in total detected OTUs) of aquatic and non-aquatic data in datasets from peer-reviewed publications (Figure 1 and Table S3). The percentage of aquatic samples, including water from hot spring, lake, river, ocean and glacier, varied from 1.87% to 62.65%, while those from other ecosystems, including various soil samples and animal skins, varied from 10.21% to 65.30% (Figure 1).

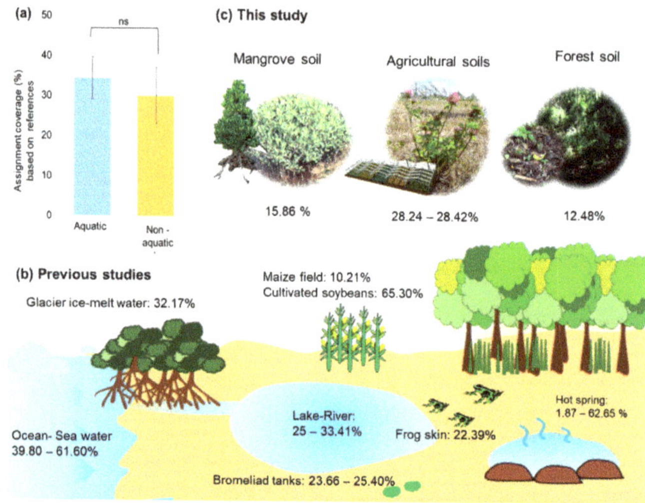


**Figure 1.** FAPROTAX assignment percentages over aquatic and non-aquatic ecosystems. (**a**) Bar plot showing the average proportion of FAPROTAX assignment based on previous studies of aquatic samples (blue) and non-aquatic samples (yellow). Error bars represent standard error of the mean and the letter "ns" indicates statistical insignificance tested by Mann-Whitney U test. (**b**) FAPROTAX assignment percentage over different sample sources based on previous literature. (**c**) The assignment percentage found in soil samples from mangrove, agriculture and forest ecosystems in this present study.

### 3.2. General Overview of Bioinformatics and FAPROTAX Assignment of Data Derived from Soil Samples

A total of 196,366 reads (on average 19,637 ± 2355 reads per sample) of bacterial 16S rRNA gene were detected in agricultural bulk soil and rhizosphere soil, after quality filtering and chimeric sequence removal. For forest and mangrove soils, 119,433 reads (on average 23,886 ± 1980 reads per sample) and 66,263 reads (on average 19,637 ± 2355 reads per sample [33]) were detected, respectively. After rare OTUs were removed, the sequences were normalized to the smallest read numbers per sample, which were 4951, 12,269 and 7086 [33] reads per sample for agricultural soils (bulk soil and rhizosphere soil), forest soil, and mangrove rhizosphere soil, respectively.

Different bacterial OTUs were obtained from agricultural bulk soil (3329), agricultural rhizosphere soil (3365), forest soil (2177) and mangrove rhizosphere soil of *R. stylosa* (2497). The proportion of OTUs that were identified to different taxonomic ranks and that of functional assignments were presented in Figure 2. Functional assignment capacities of agricultural soils (bulk soil and rhizosphere soil), forest soil and mangrove rhizosphere soil accounted for 28.24–28.42%, 12.48% and 15.86% (Figure 1) and the number of functions assigned to those samples were 34, 36, 37 and 52 functions, respectively (Figure S1). Predominant functions across all samples belonged to chemoheterotrophy, followed by aerobic chemoheterotrophy (Figure 3). However, when we focused on more specific functions, the result showed differences in dominant functions involved in biogeochemical cycling derived from each ecosystem (Figure 3).

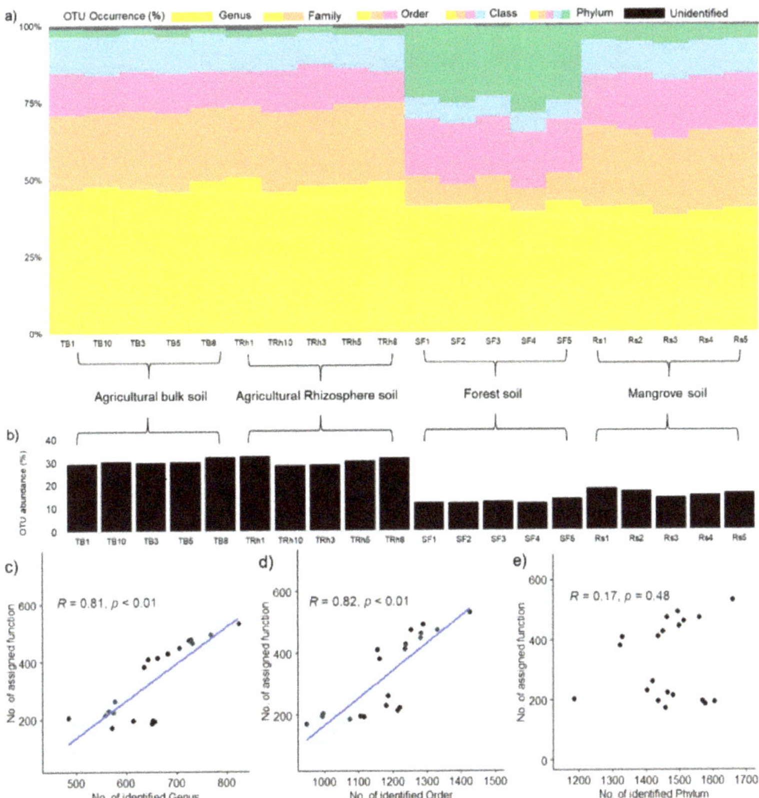


**Figure 2.** Taxonomic information across all soil samples and correlation between numbers of identified and functionally assigned OTUs. (**a**) Proportion of OTUs identifying to different taxonomic ranks. (**b**) Bar plot showing FAPROTAX's func-

tional assignment percentage (TB: agricultural bulk soil, TRh: agricultural rhizosphere soil, SF: forest soil and Rs: mangrove-rhizosphere soil). Linear regression showing Spearman's rank correlation coefficient and *p*-value on the relationship between number of FAPROTAX-functional assigned OTUs and number of identified OTUs to different taxonomic ranks which are (**c**) Genus, (**d**) Order and (**e**) Phylum. "No." is an abbreviation of "Number".

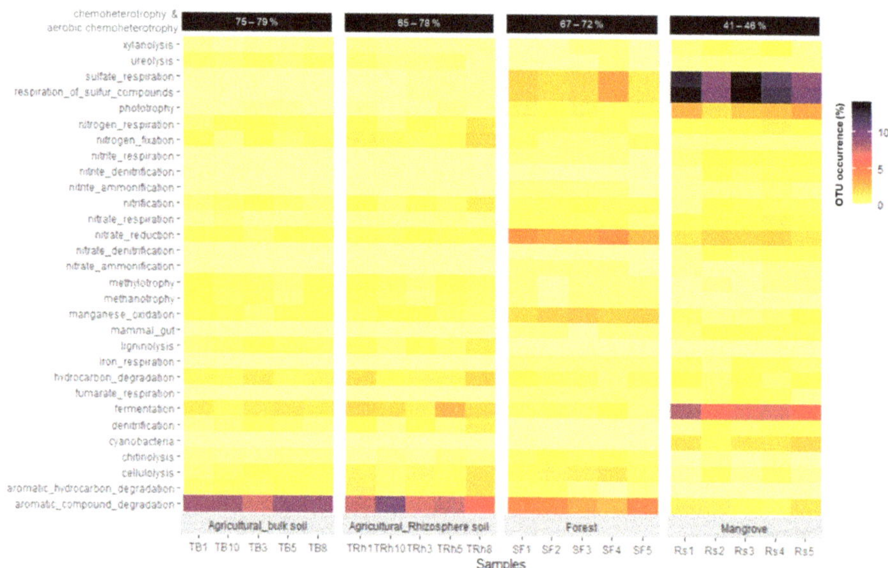


**Figure 3.** Heatmap of metabolic phenotypes/functions of bacteria. The data based on OTUs occurrence (number of OTUs capable for each function) derived from agricultural bulk soil (TB), agricultural rhizosphere soil (TRh), forest soil (SF) and mangrove-rhizosphere soil (Rs) samples.

### 3.3. Correlation between the Number of Functionally Assigned OTUs and Taxonomically Identified OTUs in Each Taxonomic Rank

Significant and positive correlations ($\sigma$ = 0.90–0.95, $p < 0.01$, Figure 2c,d) were found between the number of functionally assigned OTUs and number of OTUs assigned at genus and order levels, whereas the correlation between that at phylum level was not statistically significant ($\sigma$ = 0.11, $p$ = 0.64, Figure 2e). Although strong correlations between the numbers of functionally assigned OTUs and numbers of OTUs assigned at genus and order levels are detected in this study, we noted that a small set of data (20 samples) was used for the correlation analysis. Thus, these correlation results should be interpreted carefully.

### 3.4. Accuracy of Functionally Assigned Bacterial and Archaeal OTUs Based on FAPROTAX in Soil Systems

The result showed that more than 97% of the FAPROTAX assigned OTUs have previously been detected and potentially performed the functions in agricultural (1081 out of 1098 OTUs) and forest soils (265 out of 272 OTUs). On the other hand, only 28.79% (114 out of 396 OTUS) of functionally assigned OTUs detected in Mangrove rhizosphere soil had record of appearance and assigned functional performance in soil. We found that several detected taxa, including but not limited to *Demequina*, *Euzebya*, *Maribacter*, *Marinobacter*, *Muricauda*, *Desulfatitalea*, *Desulfopila*, were found to potentially perform the assigned functions in an estuary, seawater, sediment, and other marine habitats (Table S2.3). However, it should be noted that mangrove is a unique habitat with a mixture of land and sea, and therefore some aquatic bacteria can possibly be detected.

### 3.5. Impact of Adding Reported Datasets to Functional Assignment Percentage

In total, 17 and 31 OTUs detected in the agricultural bulk and rhizosphere soil, respectively, were functionally assigned to nitrogen fixation by FAPROTAX, while the additions of 53 and 59 OTUs were assigned the function by a manual search based on reference given in Table 1. Several taxa that could potentially perform nitrogen fixation were assigned in addition to *Rhizobium gallicum*, the only taxon that was assigned by FAPROTAX (Table 2).

**Table 2.** Number of OTUs capable of N-fixation before and after adding data from previous studies and the names of additional taxa.

| Ecosystems | No. of N-Fixation (FAPROTAX Assignment) | No. of Additional OTU Manually Assigned to N-Fixation | Additional Taxa Capable of N-Fixation |
|---|---|---|---|
| Agricultural Bulk soil | 17 | 53 | *Bacillus megaterium, Bacillus cereus, Ensifer, Bradyrhizobium, Microvirga, Phyllobacterium, Mesorhizobium, Allorhizobium-Neorhizobium-Pararhizobium-Rhizobium, Burkholderia-Caballeronia-Paraburkholderia, Devosia, Methylobacterium* |
| Agricultural Rhizosphere soil | 31 | 59 | *Bacillus megaterium, Ensifer, Bradyrhizobium, Mesorhizobium, Phyllobacterium, Microvirga, Devosia, Allorhizobium-Neorhizobium-Pararhizobium-Rhizobium, Burkholderia-Caballeronia-Paraburkholderia, Methylobacterium* |

## 4. Discussion

Keeping possible biases inherent to molecular technique and next-generation sequencing (NGS) in mind [9] (i.e., PCR, short-read sequences, pan-genome concept of bacterial evolution), many works demonstrated that NGS can be successfully used to improve our understanding of bacterial taxonomic structure and functional profile across aquatic [9,17–21] and terrestrial ecosystems [22–25]. One of the most commonly used NGS techniques (Illumina Miseq) offers paired-end reads, which increase almost two times of the non-merging read length (i.e., 580 bp for MiSeq reads of 300 bp with a 20 bp minimal overlap) [52]. In this study, we demonstrated that FAPROTAX was able to assign functions to prokaryotic taxa derived from both aquatic and terrestrial sources, especially soils. Even though aquatic samples tend to gain higher assignment than those of terrestrial samples, no significant difference was found on assignment percentage between different sample sources based on a limited number of publications (Table S3).

Several previous studies used this tool to predict the functions of bacterial/archaeal communities residing in not only aquatic habitats but also in soil samples. For example, FAPROTAX was used to assess the impact of microbial inoculation and fertilizer application on soil bacterial functions involved in for C and N cycles [53–55]. Specifically, Gao et al. [53] showed stimulation of denitrification and nitrification in soil after *Spartina alterniflora* invasion, as well as Li et al. [54] revealed a significant effect of straw incorporation and nitrogen fertilization on hydrocarbon degradation and nitrogen fixation. Similarly, Wang et al. [55] showed an increase of aerobic nitrite oxidation in soil inoculated with multi-species inoculants. Moreover, FAPROTAX was utilized in several studies that aimed to compare the effect between different treatments or a response of site managements [15,28,56–59]. For instance, it was used to investigate the effects of bacterial functions after grazing prohibition and in plantation and natural forest [15,28,56]. The increase in denitrification and nitrification functions was found after soil tillage and forest-to-agriculture conversion [57,58]. In addition to soil samples, FAPROTAX has also been helpful in predicting bacterial functions of animals and human microbiome. For examples, FAPROTAX was used to compare the functional richness of bacteria in gallbladder and gut between young and adult rabbits [60], infer biogeochemical processes, especially in nitrogen and manganese cycling occurring on human and mammalian skin [32] and show the impact of environmental factors on the functional diversity of bacteria in frog skin [14]. Although FAPROTAX is unable to reveal functional phenotypes of all taxa in the community, previous studies have shown that it was a helpful tool to highlight functions related to biogeochemical dynamics, especially on N and C cycle in non-aquatic samples and to compare functional profiles among treatments.

Our results also showed that sample source was not a primary factor that limited the application of FAPROTAX, especially in soil samples. A strong positive and significant correlation between the number of functional assigned OTUs and the number of OTUs identified to the order and genus levels, but not with those assigned to phylum level implies that poor taxonomic identification was an important factor that limited FAPROTAX functional assignment in soil samples. Since certain functions are conserved at low taxonomic levels (species, genus or order), OTUs identified only at a phylum level usually cannot be assigned to any function. The study cases on agricultural and forest soil samples reported that more than 97% of functionally assigned OTUs have previously been reported to exist (previously reported on both appearance and functional performance) in soil samples (Tables S2.1–S2.3). We found several studies that confirmed the presence of functionally assigned taxa in soil bacterial communities. For example, taxa related to C-cycle such as *Nocardioides,* which was previously isolated from soil, was assigned to aromatic compound degradation and confirmed the ability to degrade hexachlorobenzene [61]. *Rhodococcus* species was also assigned to hydrocarbons degradation and was found to have the ability to utilize various hydrocarbons groups in soil [62]. Similarly, taxa involved in nitrogen cycles, such as *Rhizobium gallicum* (nitrogen fixation) and *Micromonospora aurantiaca* (nitrate reduction and cellulolysis), were also found in soil [63,64] (see Tables S2.1–S2.3 for more information). These lines of evidence confirmed that some taxa available in the FAPROTAX database are generally dispersed across different ecosystems, including soil, even though they were originally generated from aquatic samples. Moreover, we showed that bacterial taxa and functional prediction presented in the FAPROTAX database were not only derived from aquatic samples but also from soil, plants and animals (Table S4). These results support our hypothesis that FAPROTAX was compatible with terrestrial ecosystems. On the other hand, lower portion of soil-detected taxa but high portion of marine-detected taxa in mangrove rhizosphere soil was not surprising results because mangrove land is located at the boundary between terrestrial and aquatic ecosystems. Therefore, it is possible to detect both marine and soil bacterial taxa in the area [65]. More importantly, we confirmed that a number of bacterial taxa found in mangrove rhizosphere soil were also found in soil system.

Furthermore, we demonstrated that the addition of references on bacterial taxa capable of a particular function could significantly increase the efficiency of the functional assignment. This study showed that adding more references to nitrogen fixation function can increase the number of OTUs assigned to the function by 2–3 times, accounting for a 2% increase in total functional assignment percentage. The absence of the nitrogen fixation function may be due to FAPROTAX being created based on aquatic samples [9]; thus, some terrestrial specialized taxa were neglected from the database. Furthermore, bulk and rhizosphere soils of red clover (*Trifolium pratense*) are known to colonize by diverse N-fixing bacteria [66]. Consequently, using relevant and up-to-date references of current status of N-fixing bacterial legume symbiosis can strongly increase the number of functional assigned bacterial taxa. Specifically, many important N-fixing bacterial genera, for examples *Ensifer, Bradyrhizobium, Microvirga, Mesorhizobium, Devosia,* etc., are not included in FAPROTAX as N-fixing bacteria. Therefore, FAPROTAX database still needs to be extended to cover other relevant functions in soils. This study provided soil bacterial taxa capable of nitrogen fixation, phosphate and potassium solubilization, so further study on soil sample could make the most of our work.

Although this study has demonstrated that FAPROTAX was compatible with soil samples, the bias of this tool should be kept in mind. The FAPROTAX has an assumption that if all cultured members of a taxon can perform a function, all members of that taxon (cultured and uncultured) can perform that function. Moreover, since soil ecosystem contains several biogeochemical cycles as well as diverse prokaryotic taxa, it is concerning that FAPROTAX may have low performance (low functional assignment percentage) for certain soil samples. We recommend that further work should add more soil-specific functions and prokaryotic taxa to optimize the performance of this tool. On the other hand, other alternative tools can be applied for bacterial functional annotation in soil, such as PICRUSt [11,12] or Tax4Fun [8]. Notice should be taken that each tool provides different aspects of bacterial functional phenotypes. FAPROTAX presents functional phenotypes as metabolic and ecologically relevant functions which are predicted based on the literature of cultured taxa (i.e., nitrogen fixation, nitrification, hydrocarbon degradation, chitinolysis, cellulolysis, etc.), whereas PICRUSt or Tax4Fun present functions as phenotypes of gene families or enzyme activities which are predicted based on gene content. We simulate results from Tax4Fun to show the functional phenotype of bacteria associated with the rhizosphere soil samples of *Trifolium pratense* used in this study (Appendix A). The simulated results show various potential enzymes that may be relevant to many soil functions (Figure A2). In our opinion, each functional annotation tool provides information on different aspects of bacterial functions. Selection of such tools for a particular study should depend on its purposes. If a study focuses on key bacterial functions important for biogeochemical cycling as well as microbe-microbe, plant-microbe and animal-microbe interactions, FAPROTAX can be a suitable choice. On the other hand, if a study targets at changes of gene expression or potential enzyme activity, other tools should be considered, such as PICRUSt or Tax4Fun.

In conclusion, this study presented that FAPROTAX database can be effectively used to predict function of bacteria in soil samples. Even though the database cannot predict function of all detected taxa, it can be beneficial for fast-functional screening or grouping of 16S derived bacterial data from any ecosystem. We further suggested that additional datasets of both the taxonomy and functional references could improve the FAPROTAX database and thereby increase the number of functionally assigned OTUs derived from 16S rRNA.

**Supplementary Materials:** The following are available online at https://www.mdpi.com/2076-3417/11/2/688/s1; Figure S1: Heatmap of all detected functions of bacteria across all soil samples. The data based on OTUs occurrence (number of OTUs capable for each function) derived from agricultural bulk soil (TB), agricultural rhizosphere soil (TRh), forest soil (SF) and mangrove-rhizosphere soil (Rs) samples. Table S1.1: OTU table of prokaryotic taxa detected in agricultural soil. Table S1.2: OTU table of prokaryotic taxa detected in forest soil. Table S1.3: OTU table of prokaryotic taxa detected in mangrove-rhizosphere soil. Table S2.1: FAPROTAX validation of prokaryotic taxa prokaryotic taxa detected in agricultural soils showing evident based on peer-reviewed literature of soil ecosystems on the appearance and the functional performance of prokaryotic taxa detected in agricultural soils. Table S2.2: FAPROTAX validation of prokaryotic taxa prokaryotic taxa detected in forest soil showing evident based on peer-reviewed literature of soil ecosystems on the appearance and the functional performance of prokaryotic taxa detected in forest soil. Table S2.3: FAPROTAX validation of prokaryotic taxa prokaryotic taxa detected in mangrove rhizosphere soil showing evident based on peer-reviewed literature of soil ecosystems on the appearance and the functional performance of prokaryotic taxa detected in mangrove rhizosphere soil. Table S3: Data from peer-reviewed publications showing sample sources and proportion of number of OTUs assigned by FAPROTAX to total detected OTUs. Table S4: Example of data from FAPROTAX database showing functions, prokaryotic taxa and their habitat obtained from the reference cited in FAPROTAX database.

**Author Contributions:** Conceptualization, T.D., T.W. and W.P.; field experiments, C.S., S.F.M.W., T.D. and W.P.; molecular analysis, C.S., S.F.M.W., S.H. and W.P.; bioinformatics, S.F.M.W. and T.W.; software, S.F.M.W. and T.W.; formal analysis, C.S., S.H. and W.P.; investigation, C.S., S.F.M.W., S.H. and W.P.; resources, T.D. and W.P.; data curation, C.S., S.H. and W.P.; writing—original draft preparation, C.S., W.P. and T.D.; writing—review and editing, C.S., S.F.M.W., S.H., T.D., T.W. and W.P.; visualization, C.S. and W.P.; supervision, W.P. and T.D.; funding acquisition, W.P. and T.D. All authors have read and agreed to the published version of the manuscript.

**Funding:** The next generation sequencing cost was covered by personal research budgets of W.P. from the Helmholtz Centre for Environmental Research—UFZ.

**Institutional Review Board Statement:** Not applicable.

**Informed Consent Statement:** Not applicable.

**Data Availability Statement:** Publicly available datasets were analyzed in this study. This data can be found under BioProject accession number: PRJNA646011 and PRJNA554586.

**Acknowledgments:** We appreciate the Helmholtz Association, the Federal Ministry of Education and Research, the State Ministry of Science and Economy of Saxony-Anhalt, and the State Ministry for Higher Education, Research and the Arts Saxony to fund the Global Change Experimental Facility (GCEF) project. We thank to Martin Schädler for his role in setting up and maintaining the GCEF project and the staffs of the Bad Lauchstädt Experimental Research Station for their work in maintaining the plots and infrastructures of the GCEF. Benjawan Tanunchai and Dolaya Sadubsarn are thanks for helping with field experiments and molecular works. We also thank to Forest Restoration Research Unit (FORRU-CMU) for their support on field experiment in Thailand. C.S. thanks to Science Achievement Scholarship of Thailand for financial support. S.F.M.W. is supported financially by the Egyptian Scholarship (Ministry of Higher Education, External Missions 2016/2017 call). T.D. is partially supported by Chiang Mai University.

**Conflicts of Interest:** The authors declare no conflict of interest.

## Appendix A. Functional Phenotypes Derived from Tax4Fun and FAPROTAX. A Simulate Result from Bacterial Communities of Agricultural Rhizosphere of *Trifolium pretense*

A dataset (OTU table) consisted of bacterial communities in the rhizosphere of *Trifolium pratense* was selected to simulate the result derived from Tax4Fun. In detail, the Tax4Fun [8] R package, which employs 16S rRNA gene-based taxonomic information, and the Kyoto Encyclopedia of Genes and Genomes (KEGG) database were used to predict the metabolic functional attributes of bacterial communities in the rhizosphere of *T. pratense*. Tax4Fun converted the SILVA-labelled OTUs into prokaryotic KEGG organisms and normalized these predictions using the 16S rRNA copy number (obtained from the National Center for Biotechnology Information genome annotations). Furthermore, FAPROTAX was also used to predicted bacterial function of the same dataset following the FAPROTAX's instruction (http://www.loucalab.com/archive/FAPROTAX/) [9].

A total of 36 functions were derived from FAPROTAX (Figure A1). FAPROTAX showed ecologically relevant functions which concern biogeological cycles, such as nitrification, cellulolysis and hydrocarbon degradation, and the interaction of microbe to plants/animals, such as plant pathogen or invertebrate parasites (Figure A1). On the other hand, Tax4Fun provided a total of 6338 functions which included functions that involved in soil biogeochemical cycles, such as chitinase, beta-glucosidase or acid phosphatase, and not involved in soil system, such as sn-glycerol 3-phosphate transport system ATP-binding protein, queuine tRNA-ribosyltransferase, and DNA-directed RNA polymerase. In order to use Tax4Fun, the researcher might have an overview of functions required for their work, then sort only interesting functions to be investigated. For example, in this case, we gathered 30 enzymes involved in soil system (Table A1) and show the functional profile presented in Figure A2.

However, we believe that both functional annotation tool contributes most benefit, but different aspect for each work. The selection of these tools should depend on the purpose of each study. If a study focuses on the biogeochemical cycle or the interaction of microbe to plants/animals, FAPROTAX would be the best alternatives. Moreover, FAPRROTAX is quite easy to follow as it provides plain functional phenotypes. On the other hand, if a study pays more attention to enzyme activity, Tax4Fun should be a great choice.

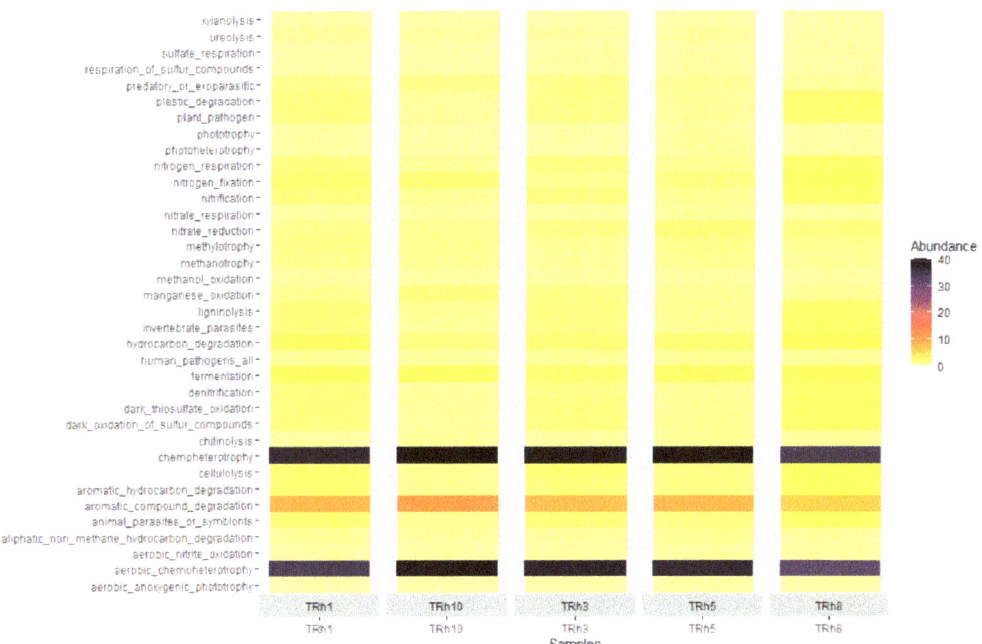


**Figure A1.** Heatmap shows all functions of bacteria predicted by FAPROTAX. The data based on OTUs abundance derived from rhizosphere soil samples (TRh) of red clover (*Trifolium pratense*).

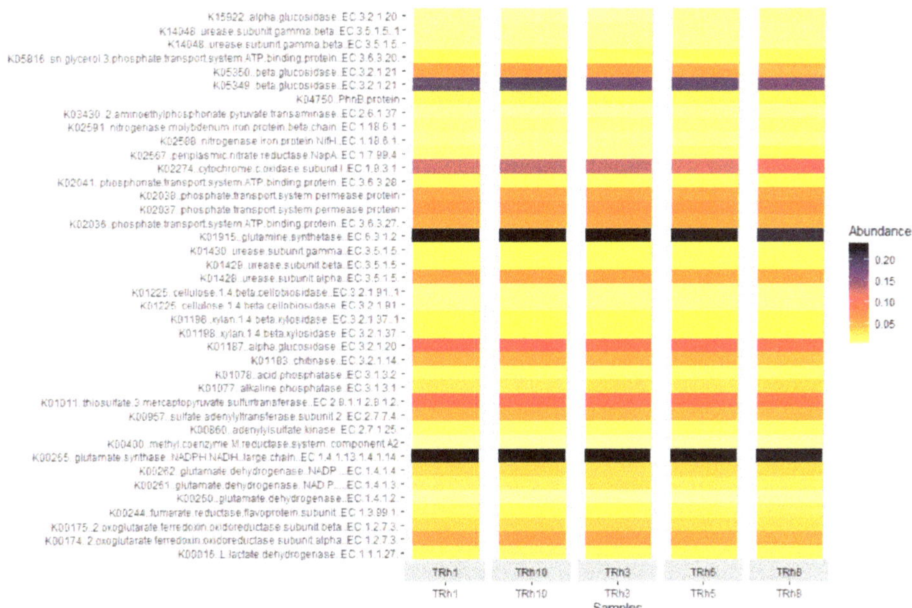


**Figure A2.** Heatmap shows functions of bacteria predicted by Tax4Fun. The data based on OTUs abundance derived from rhizosphere soil samples (TRh) of red clover (*Trifolium pratense*).

**Table A1.** Tax4Func functions involved in soil system and its description.

| Tax4Fun Output | Function's Description | References |
|---|---|---|
| K02274; cytochrome c oxidase subunit I [EC:1.9.3.1] | Aerobic Respiration | [67] |
| K00860; adenylylsulfate kinase [EC:2.7.1.25] | Assimilatory Sulfate Reduction | [67] |
| K00957; sulfate adenylyltransferase subunit 2 [EC:2.7.7.4] | Assimilatory Sulfate Reduction | [67] |
| K00016; L-lactate dehydrogenase [EC:1.1.1.27] | Fermentation | [67] |
| K00400; methyl coenzyme M reductase system, component A2 | Methanogenesis | [67] |
| K00265; glutamate synthase (NADPH/NADH) large chain [EC:1.4.1.13 1.4.1.14] | Nitrogen Assimilation | [67] |
| K01915; glutamine synthetase [EC:6.3.1.2] | Nitrogen Assimilation | [67] |
| K02588; nitrogenase iron protein NifH [EC:1.18.6.1] | Nitrogen Fixation | [67] |
| K02591; nitrogenase molybdenum-iron protein beta chain [EC:1.18.6.1] | Nitrogen Fixation | [67] |
| K00261; glutamate dehydrogenase (NAD(P)+) [EC:1.4.1.3] | Nitrogen Mineralization | [67] |
| K00262; glutamate dehydrogenase (NADP+) [EC:1.4.1.4] | Nitrogen Mineralization | [67] |
| K00260; glutamate dehydrogenase [EC:1.4.1.2] | Nitrogen Mineralization | [67] |
| K02567; periplasmic nitrate reductase NapA [EC:1.7.99.4] | Nitrogen Reduction | [67] |
| K01011; thiosulfate/3-mercaptopyruvate sulfurtransferase [EC:2.8.1.1 2.8.1.2] | Sulfur Mineralisation | [67] |
| K01077; alkaline phosphatase [EC:3.1.3.1] | Cleaving of PO4 from P-containing OM | [68] |
| K01078; acid phosphatase [EC:3.1.3.2] | Cleaving of PO4 from P-containing OM | [68] |
| K01183; chitinase [EC:3.2.1.14] | Hydrolysis of chitooligosaccharides | [68] |
| K01225; cellulose 1,4-beta-cellobiosidase [EC:3.2.1.91] | Hydrolysis of cellulose | [68] |
| K01198; xylan 1,4-beta-xylosidase [EC:3.2.1.37] | Hydrolysis of hemicellulose | [68] |
| K05349; beta-glucosidase [EC:3.2.1.21] | Hydrolysis of cellulose | [68] |
| K05350; beta-glucosidase [EC:3.2.1.21] | Hydrolysis of cellulose | [68] |
| K14048; urease subunit gamma/beta [EC:3.5.1.5] | Hydrolysis of urea | [68] |
| K01428; urease subunit alpha [EC:3.5.1.5] | Hydrolysis of urea | [68] |
| K01429; urease subunit beta [EC:3.5.1.5] | Hydrolysis of urea | [68] |
| K01430; urease subunit gamma [EC:3.5.1.5] | Hydrolysis of urea | [68] |
| K14048; urease subunit gamma/beta [EC:3.5.1.5] | Hydrolysis of urea | [68] |
| K15922; alpha-glucosidase [EC:3.2.1.20] | Hydrolysis of soluble saccharides | [68] |
| K01187; alpha-glucosidase [EC:3.2.1.20] | Hydrolysis of soluble saccharides | [68] |
| K01198; xylan 1,4-beta-xylosidase [EC:3.2.1.37] | Release of xylose from short xylan oligomers | [69] |
| K01225; cellulose 1,4-beta-cellobiosidase [EC:3.2.1.91] | Release of cellobiose from non-reducing end of cellulose chains | [69] |

## References


1. Graham, E.B.; Knelman, J.E.; Schindlbacher, A.; Siciliano, S.; Breulmann, M.; Yannarell, A.; Beman, J.M.; Abell, G.; Philippot, L.; Prosser, J.; et al. Microbes as Engines of Ecosystem Function: When Does Community Structure Enhance Predictions of Ecosystem Processes? *Front. Microbiol.* **2016**, *7*, 214. [CrossRef] [PubMed]
2. Seo, J.-S.; Keum, Y.S.; Li, Q.X. Bacterial Degradation of Aromatic Compounds. *Int. J. Environ. Res. Public Health* **2009**, *6*, 278–309. [CrossRef] [PubMed]
3. Jahangeer, J.; Kumar, V. An Overview on Microbial Degradation of Petroleum Hydrocarbon Contaminants. *Int. J. Eng. Tech. Res.* **2013**, *1*, 34–37.
4. Igiri, B.E.; Okoduwa, S.I.R.; Idoko, G.O.; Akabuogu, E.P.; Adeyi, A.O.; Ejiogu, I.K. Toxicity and Bioremediation of Heavy Metals Contaminated Ecosystem from Tannery Wastewater: A Review. *J. Toxicol.* **2018**, *2018*, 2568038. [CrossRef] [PubMed]
5. White, C.; Shaman, A.K.; Gadd, G.M. An Integrated Microbial Process for the Bioremediation of Soil Contaminated with Toxic Metals. *Nat. Biotechnol.* **1998**, *16*, 572–575. [CrossRef] [PubMed]
6. Sitte, J.; Akob, D.M.; Kaufmann, C.; Finster, K.; Banerjee, D.; Burkhardt, E.-M.; Kostka, J.E.; Scheinost, A.C.; Büchel, G.; Küsel, K. Microbial Links between Sulfate Reduction and Metal Retention in Uranium- and Heavy Metal-Contaminated Soil. *Appl. Environ. Microbiol.* **2010**, *76*, 3143–3152. [CrossRef] [PubMed]
7. Gómez-Ramírez, M.; Zarco-Tovar, K.; Aburto, J.; De León, R.G.; Rojas-Avelizapa, N.G. Microbial Treatment of Sulfur-Contaminated Industrial Wastes. *J. Environ. Sci. Health Part A* **2014**, *49*, 228–232. [CrossRef]

8. Aßhauer, K.P.; Wemheuer, B.; Daniel, R.; Meinicke, P. Tax4Fun: Predicting Functional Profiles from Metagenomic 16S rRNA Data. *Bioinformatics* **2015**, *31*, 2882–2884. [CrossRef]
9. Louca, S.; Parfrey, L.W.; Doebeli, M. Decoupling Function and Taxonomy in the Global Ocean Microbiome. *Science* **2016**, *353*, 1272–1277. [CrossRef]
10. Nguyen, N.H.; Song, Z.; Bates, S.T.; Branco, S.; Tedersoo, L.; Menke, J.; Schilling, J.S.; Kennedy, P.G. FUNGuild: An Open Annotation Tool for Parsing Fungal Community Datasets by Ecological Guild. *Fungal Ecol.* **2016**, *20*, 241–248. [CrossRef]
11. Langille, M.G.I.; Zaneveld, J.; Beiko, R.G.; Huttenhower, C.; Caporaso, J.G.; McDonald, D.; Knights, D.; Reyes, J.A.; Clemente, J.C.; Burkepile, D.E.; et al. Predictive Functional Profiling of Microbial Communities Using 16S rRNA Marker Gene Sequences. *Nat. Biotechnol.* **2013**, *31*, 814–821. [CrossRef] [PubMed]
12. Douglas, G.M.; Maffei, V.J.; Zaneveld, J.; Yurgel, S.N.; Brown, J.R.; Taylor, C.M.; Huttenhower, C.; Langille, M.G.I. PICRUSt2: An improved and customizable approach for metagenome inference. *BioRxiv* **2020**. [CrossRef]
13. Bowman, J.S.; Ducklow, H.W. Microbial Communities Can Be Described by Metabolic Structure: A General Framework and Application to a Seasonally Variable, Depth-Stratified Microbial Community from the Coastal West Antarctic Peninsula. *PLoS ONE* **2015**, *10*, e0135868. [CrossRef] [PubMed]
14. Varela, B.J.; Lesbarrères, D.; Ibáñez, R.; Green, D.M. Environmental and Host Effects on Skin Bacterial Community Composition in Panamanian Frogs. *Front. Microbiol.* **2018**, *9*, 298. [CrossRef] [PubMed]
15. Deng, J.; Zhang, Y.; Yin, Y.; Zhu, X.; Zhu, W.; Zhou, Y. Comparison of Soil Bacterial Community and Functional Characteristics Following Afforestation in the Semi-Arid Areas. *PeerJ* **2019**, *7*, e7141. [CrossRef]
16. Wilmotte, A.; Herdman, M. *Bergey's Manual of Systematic Bacteriology: Volume One: The Archaea and the Deeply Branching and Phototrophic Bacteria*, 2nd ed.; Garrity, G., Boone, D.R., Castenholz, R.W., Eds.; Springer: New York, NY, USA, 2001; ISBN 978-0-387-98771-2.
17. Scheutz, F.; Strockbine, N.A.; Genus, I. *Bergey's Manual of Systematic Bacteriology: Volume 2: The Proteobacteria*, 2nd ed.; Springer: New York, NY, USA, 2005; ISBN 978-0-387-95040-2.
18. Vos, P.; Garrity, G.; Jones, D.; Krieg, N.R.; Ludwig, W.; Rainey, F.A.; Schleifer, K.H.; Whitman, W.B. (Eds.) *Bergey's Manual of Systematic Bacteriology: Volume 3: The Firmicutes*, 2nd ed.; Springer: New York, NY, USA, 2009; ISBN 978-0-387-95041-9.
19. Whitman, W.B. *Bergey's Manual of Systematic Bacteriology: Volume 4: The Bacteroidetes, Spirochaetes, Tenericutes (Mollicutes), Acidobacteria, Fibrobacteres, Fusobacteria, Dictyoglomi, Gemmatimonadetes, Lentisphaerae, Verrucomicrobia, Chlamydiae, and Planctomycetes*, 2nd ed.; Krieg, N.R., Ludwig, W., Whitman, W., Hedlund, B.P., Paster, B.J., Staley, J.T., Ward, N., Brown, D., Parte, A., Eds.; Springer: New York, NY, USA, 2010; ISBN 978-0-387-95042-6.
20. Whitman, W.; Goodfellow, M.; Kämpfer, P.; Busse, H.-J.; Trujillo, M.; Ludwig, W.; Suzuki, K.; Parte, A. (Eds.) *Bergey's Manual of Systematic Bacteriology: Volume 5: The Actinobacteria*, 2nd ed.; Springer: New York, NY, USA, 2012; ISBN 978-0-387-95043-3.
21. Lory, S. *The Prokaryotes: Prokaryotic Biology and Symbiotic Associations*; Rosenberg, E., DeLong, E.F., Lory, S., Stackebrandt, E., Thompson, F., Eds.; Springer: Berlin/Heidelberg, Germany, 2013; pp. 359–400, ISBN 978-3-642-30194-0.
22. International Journal of Systematic and Evolutionary Microbiology. Available online: https://www.microbiologyresearch.org/content/journal/ijsem (accessed on 24 December 2020).
23. Louca, S.; Jacques, S.M.S.; Pires, A.P.F.; Leal, J.S.; Srivastava, D.S.; Parfrey, L.W.; Farjalla, V.F.; Doebeli, M. High Taxonomic Variability despite Stable Functional Structure Across Microbial Communities. *Nat. Ecol. Evol.* **2017**, *1*, 1–12. [CrossRef]
24. Hu, A.; Li, S.; Zhang, L.; Wang, H.; Yang, J.R.; Luo, Z.; Rashid, A.; Chen, S.; Huang, W.; Yu, C.-P. Prokaryotic Footprints in Urban Water Ecosystems: A Case Study of Urban Landscape Ponds in a Coastal City, China. *Environ. Pollut.* **2018**, *242*, 1729–1739. [CrossRef]
25. Zhang, Y.; Wu, G.; Jiang, H.; Yang, J.; She, W.; Khan, I.; Li, W. Abundant and Rare Microbial Biospheres Respond Differently to Environmental and Spatial Factors in Tibetan Hot Springs. *Front. Microbiol.* **2018**, *9*, 2096. [CrossRef]
26. Bomberg, M.; Liljedahl, L.C.; Lamminmäki, T.; Kontula, A. Highly Diverse Aquatic Microbial Communities Separated by Permafrost in Greenland Show Distinct Features According to Environmental Niches. *Front. Microbiol.* **2019**, *10*, 1583. [CrossRef]
27. Yang, Y.; Hou, Y.; Ma, M.; Zhan, A. Potential Pathogen Communities in Highly Polluted River Ecosystems: Geographical Distribution and Environmental Influence. *Ambio* **2019**, *49*, 197–207. [CrossRef]
28. Wei, H.; Peng, C.; Liu, X.; Chen, D.; Li, Y.; Wang, M.; Yang, B.; Song, H.; Li, Q.; Jiang, L.; et al. Contrasting Soil Bacterial Community, Diversity, and Function in Two Forests in China. *Front. Microbiol.* **2018**, *9*, 1693. [CrossRef] [PubMed]
29. Chang, C.; Chen, W.; Luo, S.; Ma, L.; Li, X.; Tian, C. Rhizosphere Microbiota Assemblage Associated with Wild and Cultivated Soybeans Grown in Three Types of Soil Suspensions. *Arch. Agron. Soil Sci.* **2018**, *65*, 74–87. [CrossRef]
30. Khan, M.A.W.; Bohannan, B.J.M.; Nüsslein, K.; Tiedje, J.M.; Tringe, S.G.; Parlade, E.; Barberán, A.; Rodrigues, J.L.M. Deforestation Impacts Network co-occurrence Patterns of Microbial Communities in Amazon Soils. *FEMS Microbiol. Ecol.* **2019**, *95*. [CrossRef] [PubMed]
31. Shen, C.; Ma, D.; Sun, R.; Zhang, B.; Li, D.; Ge, Y. Long-term Stacking Coal Promoted Soil Bacterial Richness Associated with Increased Soil Organic Matter in Coal Yards of Power Plants. *J. Soils Sediments* **2019**, *19*, 3442–3452. [CrossRef]
32. Stegelmeier, A.A.; Müller, K.M.; Weese, J.S.; Neufeld, J.D. Comprehensive Skin Microbiome Analysis Reveals the Uniqueness of Human Skin and Evidence for Phylosymbiosis within the Class Mammalia. *Proc. Natl. Acad. Sci. USA* **2018**, *115*, E5786–E5795. [CrossRef]

33. Purahong, W.; Sadubsarn, D.; Tanunchai, B.; Wahdan, S.F.M.; Sansupa, C.; Noll, M.; Lei, B.; Buscot, F.; Wu, N. First Insights into the Microbiome of a Mangrove Tree Reveal Significant Differences in Taxonomic and Functional Composition among Plant and Soil Compartments. *Microorganisms* **2019**, *7*, 585. [CrossRef] [PubMed]
34. Kathiresan, K.; Bingham, B.L. Biology of mangroves and mangrove Ecosystems. *Adv. Mar. Biol.* **2001**, *40*, 81–251. [CrossRef]
35. Schädler, M.; Buscot, F.; Schulz, E.; Auge, H.; Klotz, S.; Reitz, T.; Durka, W.; Bumberger, J.; Merbach, I.; Michalski, S.G.; et al. Investigating the Consequences of Climate Change under Different Land-Use Regimes: A Novel Experimental Infrastructure. *Ecosphere* **2019**, *10*, e02635. [CrossRef]
36. Wahdan, S.F.M.; Hossen, S.; Tanunchai, B.; Schädler, M.; Buscot, F.; Purahong, W. Future Climate Significantly Alters Fungal Plant Pathogen Dynamics during the Early Phase of Wheat Litter Decomposition. *Microorganisms* **2020**, *8*, 908. [CrossRef]
37. Barillot, C.D.C.; Sarde, C.-O.; Bert, V.; Tarnaud, E.; Cochet, N. A standardized method for the sampling of rhizosphere and rhizoplan soil bacteria associated to a herbaceous root system. *Ann. Microbiol.* **2012**, *63*, 471–476. [CrossRef]
38. Maxwell, J.F.; Elliott, S. *Vegetation and Vascular Flora of Doi Sutep-Pui National Park, Northern Thailand*; Biodiversity Research and Training Program: Bangkok, Thailand, 2001.
39. Klindworth, A.; Pruesse, E.; Schweer, T.; Peplies, J.; Quast, C.; Horn, M.; Glöckner, F.O. Evaluation of General 16S Ribosomal RNA Gene PCR Primers for Classical and Next-Generation Sequencing-Based Diversity Studies. *Nucleic Acids Res.* **2012**, *41*, e1. [CrossRef] [PubMed]
40. Schloss, P.D.; Westcott, S.L.; Sahl, J.W.; Stres, B.; Thallinger, G.G.; Van Horn, D.J.; Weber, C.F.; Ryabin, T.; Hall, J.R.; Hartmann, M.; et al. Introducing Mothur: Open-Source, Platform-Independent, Community-Supported Software for Describing and Comparing Microbial Communities. *Appl. Environ. Microbiol.* **2009**, *75*, 7537–7541. [CrossRef] [PubMed]
41. Edgar, R.C.; Haas, B.J.; Clemente, J.C.; Quince, C.; Knight, R. UCHIME Improves Sensitivity and Speed of Chimera Detection. *Bioinformatics* **2011**, *27*, 2194–2200. [CrossRef] [PubMed]
42. Pruesse, E.; Quast, C.; Knittel, K.; Fuchs, B.M.; Ludwig, W.; Peplies, J.; Glöckner, F.O. SILVA: A comprehensive online resource for quality checked and aligned ribosomal RNA sequence data compatible with ARB. *Nucleic Acids Res.* **2007**, *35*, 7188–7196. [CrossRef] [PubMed]
43. Hammer, Ø.; Harper, D.A.T.; Ryan, P.D. PAST: Paleontological Statistical Software Package for Education and Data Analysis. *Palaeontol. Electron.* **2001**, *4*, 1–9.
44. Kassambara, A. Ggpubr: "ggplot2" Based Publication Ready Plots. 2020. Available online: https://cran.r-project.org/web/packages/ggpubr/index.html (accessed on 17 June 2020).
45. R Core Team. 2019. Available online: https://www.eea.europa.eu/data-and-maps/indicators/oxygen-consuming-substances-in-rivers/r-development-core-team-2006 (accessed on 17 June 2020).
46. Velázquez, E.; García-Fraile, P.; Ramírez-Bahena, M.H.; Rivas, R.; Martínez-Molina, E. Current Status of the Taxonomy of Bacteria Able to Establish Nitrogen-Fixing Legume Symbiosis. In *Microbes for Legume Improvement*; Springer Science and Business Media LLC: Cham, Switzerland, 2017; pp. 1–43.
47. Yousuf, J.; Jabir, T.; Rahiman, M.; Krishnankutty, S.; Alikunj, A.P.; Abdulla, M.H.A. Nitrogen Fixing Potential of Various Heterotrophic Bacillus Strains from a Tropical Estuary and Adjacent Coastal Regions. *J. Basic Microbiol.* **2017**, *57*, 922–932. [CrossRef]
48. Zarjani, J.K.; Aliasgharzad, N.; Oustan, S.; Emadi, M.; Ahmadi, A. Isolation and Characterization of Potassium Solubilizing Bacteria in some Iranian Soils. *Arch. Agron. Soil Sci.* **2013**, *59*, 1713–1723. [CrossRef]
49. Etesami, H.; Emami, S.; Alikhani, H.A. Potassium Solubilizing Bacteria (KSB): Mechanisms, Promotion of Plant Growth, and Future Prospects A Review. *J. Soil Sci. Plant Nutr.* **2017**, *17*, 897–911. [CrossRef]
50. Sharma, S.B.; Sayyed, R.Z.; Trivedi, M.H.; Gobi, A.T. Phosphate Solubilizing Microbes: Sustainable Approach for Managing Phosphorus Deficiency in Agricultural Soils. *SpringerPlus* **2013**, *2*, 1–14. [CrossRef]
51. Kalayu, G. Phosphate Solubilizing Microorganisms: Promising Approach as Biofertilizers. *Int. J. Agron.* **2019**, *2019*, 1–7. [CrossRef]
52. Liu, T.; Chen, C.-Y.; Chen-Deng, A.; Chen, Y.-L.; Wang, J.Y.; Hou, Y.-I.; Lin, M.-C. Joining Illumina Paired-End Reads for Classifying Phylogenetic Marker Sequences. *BMC Bioinform.* **2020**, *21*, 1–13. [CrossRef] [PubMed]
53. Gao, G.-F.; Li, P.-F.; Gao, C.-H.; Zheng, H.-L.; Zhong, J.-X.; Shen, Z.-J.; Chen, J.; Li, Y.-T.; Isabwe, A.; Zhu, X.-Y.; et al. Spartina Alterniflora Invasion Alters Soil Bacterial Communities and Enhances Soil $N_2O$ Emissions by Stimulating Soil Denitrification in Mangrove Wetland. *Sci. Total Environ.* **2019**, *653*, 231–240. [CrossRef] [PubMed]
54. Li, H.; Zhang, Y.; Yang, S.; Wang, Z.; Feng, X.; Liu, H.; Jiang, Y. Variations in soil bacterial taxonomic profiles and putative functions in response to straw incorporation combined with N fertilization during the maize growing season. *Agric. Ecosyst. Environ.* **2019**, *283*, 106578. [CrossRef]
55. Wang, J.; Li, Q.; Xu, S.; Zhao, W.; Lei, Y.; Song, C.; Huang, Z. Traits-Based Integration of Multi-Species Inoculants Facilitates Shifts of Indigenous Soil Bacterial Community. *Front. Microbiol.* **2018**, *9*, 1692. [CrossRef]
56. Yin, Y.; Wang, Y.; Li, S.; Liu, Y.; Zhao, W.; Ma, Y.; Bao, G. Soil Microbial Character Response to Plant Community Variation after Grazing Prohibition for 10 Years in a Qinghai-Tibetan Alpine Meadow. *Plant Soil* **2019**, 1–15. [CrossRef]
57. Legrand, F.; Picot, A.; Cobo-Díaz, J.F.; Carof, M.; Le Floch, G. Effect of Tillage and Static Abiotic Soil Properties on Microbial Diversity. *Appl. Soil Ecol.* **2018**, *132*, 135–145. [CrossRef]
58. Merloti, L.F.; Mendes, L.W.; Pedrinho, A.; De Souza, L.F.; Ferrari, B.M.; Tsai, S.M. Forest-to-Agriculture Conversion in Amazon Drives Soil Microbial Communities and N-cycle. *Soil Biol. Biochem.* **2019**, *137*. [CrossRef]

59. Sengupta, A.; Kushwaha, P.; Jim, A.; Troch, P.; Maier, R.M. New Soil, Old Plants, and Ubiquitous Microbes: Evaluating the Potential of Incipient Basaltic Soil to Support Native Plant Growth and Influence Belowground Soil Microbial Community Composition. *Sustainability* **2020**, *12*, 4209. [CrossRef]
60. Xing, Y.; Liu, J.; Lu, F.; Wang, L.; Li, Y.; Ouyang, C. Dynamic Distribution of Gallbladder Microbiota in Rabbit at Different Ages and Health States. *PLoS ONE* **2019**, *14*, e0211828. [CrossRef]
61. Takagi, K.; Iwasaki, A.; Kamei, I.; Satsuma, K.; Yoshioka, Y.; Harada, N. Aerobic Mineralization of Hexachlorobenzene by Newly Isolated Pentachloronitrobenzene-Degrading *Nocardioides* sp. Strain PD653. *Appl. Environ. Microbiol.* **2009**, *75*, 4452–4458. [CrossRef]
62. Far, S.; Rasti, A.; Riahi, M.A. Investigation of *Rhodococcus Equi* Effects on Crude Oil from Biological Degradation Aspects by SARA, FT-IR and GC Technique. *Am. J. Petrochem.* **2019**, *1*, 1–5.
63. Mnasri, B.; Tajini, F.; Trabelsi, M.; Aouani, M.E.; Mhamdi, R. *Rhizobium Gallicum* as an Efficient Symbiont for Bean Cultivation. *Agron. Sustain. Dev.* **2007**, *27*, 331–336. [CrossRef]
64. Thawai, C.; Tanasupawat, S.; Itoh, T.; Suwanborirux, K.; Kudo, T. *Micromonospora Aurantionigra* sp. nov., Isolated from a Peat Swamp Forest in Thailand. *Actinomycetologica* **2004**, *18*, 8–14. [CrossRef]
65. Thatoi, H.; Behera, B.C.; Mishra, R.R.; Dutta, S.K. Biodiversity and Biotechnological Potential of Microorganisms from Mangrove Ecosystems: A Review. *Ann. Microbiol.* **2013**, *63*, 1–19. [CrossRef]
66. Hartman, K.; Van Der Heijden, M.G.; Roussely-Provent, V.; Walser, J.-C.; Schlaeppi, K. Deciphering Composition and Function of the Root Microbiome of a Legume Plant. *Microbiome* **2017**, *5*, 1–13. [CrossRef]
67. Caradonia, F.; Ronga, D.; Catellani, M.; Giaretta Azevedo, C.V.; Terrazas, R.A.; Robertson-Albertyn, S.; Francia, E.; Bulgarelli, D. Nitrogen Fertilizers Shape the Composition and Predicted Functions of the Microbiota of Field-Grown Tomato Plants. *Phytobiomes J.* **2019**, *3*, 315–325. [CrossRef]

68. Jian, S.; Li, J.; Chen, J.; Wang, G.; Mayes, M.A.; Dzantor, K.E.; Hui, D.; Luo, Y. Soil Extracellular Enzyme Activities, Soil Carbon and Nitrogen Storage under Nitrogen Fertilization: A Meta-Analysis. *Soil Biol. Biochem.* **2016**, *101*, 32–43. [CrossRef]
69. Weintraub, S.R.; Wieder, W.R.; Cleveland, C.C.; Townsend, A.R. Organic Matter Inputs Shift Soil Enzyme Activity and Allocation Patterns in a Wet Tropical Forest. *Biogeochemistry* **2013**, *114*, 313–326. [CrossRef]

MDPI

*Review*

# Methods for Studying Bacterial–Fungal Interactions in the Microenvironments of Soil

**Edoardo Mandolini *, Maraike Probst and Ursula Peintner ***

Institute of Microbiology, University of Innsbruck, Technikerstrasse 25, 6020 Innsbruck, Austria; Maraike.Probst@uibk.ac.at
* Correspondence: Edoardo.Mandolini@uibk.ac.at (E.M.); Ursula.Peitner@uibk.ac.at (U.P.)

**Abstract:** Due to their small size, microorganisms directly experience only a tiny portion of the environmental heterogeneity manifested in the soil. The microscale variations in soil properties constrain the distribution of fungi and bacteria, and the extent to which they can interact with each other, thereby directly influencing their behavior and ecological roles. Thus, to obtain a realistic understanding of bacterial–fungal interactions, the spatiotemporal complexity of their microenvironments must be accounted for. The objective of this review is to further raise awareness of this important aspect and to discuss an overview of possible methodologies, some of easier applicability than others, that can be implemented in the experimental design in this field of research. The experimental design can be rationalized in three different scales, namely reconstructing the physicochemical complexity of the soil matrix, identifying and locating fungi and bacteria to depict their physical interactions, and, lastly, analyzing their molecular environment to describe their activity. In the long term, only relevant experimental data at the cell-to-cell level can provide the base for any solid theory or model that may serve for accurate functional prediction at the ecosystem level. The way to this level of application is still long, but we should all start small.

**Keywords:** soil microbiology; cultivation; isotope probing; nanoSIMS; microfluidics; heterogeneity; microbial communication; distribution

**Citation:** Mandolini, E.; Probst, M.; Peintner, U. Methods for Studying Bacterial–Fungal Interactions in the Microenvironments of Soil. *Appl. Sci.* **2021**, *11*, 9182. https://doi.org/10.3390/app11199182

Academic Editor: Oleh Andrukhov

Received: 7 August 2021
Accepted: 30 September 2021
Published: 2 October 2021

**Publisher's Note:** MDPI stays neutral with regard to jurisdictional claims in published maps and institutional affiliations.



## 1. Introduction

Living organisms are always and constantly interacting with their biotic and abiotic environment, irrespective of habitat, trophic level, or biological function. The network of interacting organisms is a mosaic and entwined reality that, depending on the scale of analysis, may be described at different levels of complexity. Cumulative cell-to-cell interactions among organisms belonging to very different taxa levels and origins, such as animals, protists, fungi, bacteria, archaea, and viruses, determine the overall microbial community activity in a given habitat. These interactions have an effect not just on their surrounding environment (i.e., microenvironment), but also influence large-scale fluxes and, thus, impact global ecosystem processes [1]. However, especially for soil biota, there is still a marked gap between studies performed in laboratory conditions and the in vivo reality [2]. Although providing crucially relevant data, this reduction in complexity cannot resolve general principles of microbial interactions at the ecosystem level [3].

In this review, we focus on bacterial and fungal interactions (BFI) in soil, and how the technologies available today can be applied for a better understanding of the behavior and phenotypic diversity of these organisms when closely linked to their micro-habitat. Information on the properties of soil and on how its spatial and chemical heterogeneity affects the dynamics of BFI will be described first, to justify the selection of strategies and methodologies reported in the later sections. This minireview is not meant to be exhaustive on either topic, as the arrays of approaches and studies available is extensive and beyond the means of this review. When possible, relevant examples on BFI are given but, in many cases, it was inevitable to refer to research on interactions within bacteria or other taxa.

## 2. Fungi, Bacteria and Their Microenvironment

Interactions within the soil biota are an especially challenging and fascinating topic; as a consequence of the microorganisms' size, the variable metabolic activities of nearby microbes, and the changes in physicochemical conditions over short intervals of time and distance (soil heterogeneity), numerous microenvironments can co-exist in close proximity within a given habitat. Understanding the fine scale heterogeneity of soil environments is a prerequisite for predicting and contextualizing the physiology of the present organisms and metabolic interactions among community members. Their dynamics, in all their shapes and forms, are the consequences of their adaptation in response to the micro-habitats they experience (Figure 1).

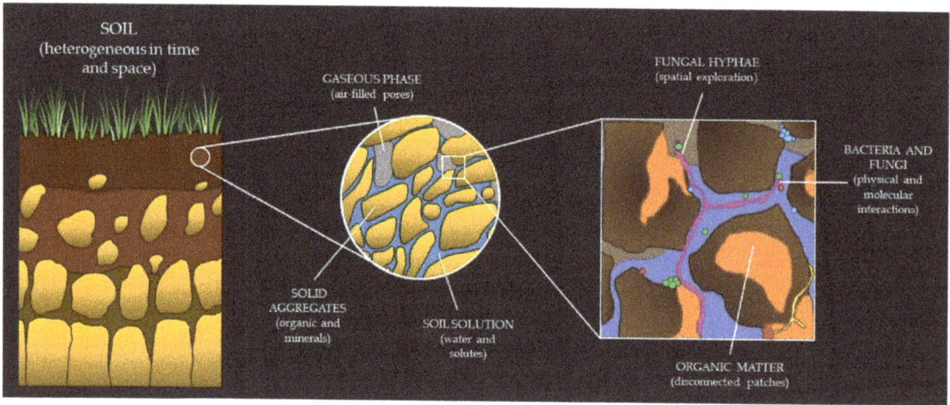


**Figure 1.** The complexity of soil depends on the scale of analysis and is ultimately a mosaic of dynamic microenvironments defined at a specific point in time and space. Fungi and bacteria, together with other soil residents, live, grow and interact in this reality to the best of their fitness, perpetually looking for their *realized niche*.

### *2.1. The Hidden Properties of Soil*

Basic physical laws govern and dictate the properties of each of the microscopic components that compose soils; the surface-to-volume ratio increases with the decreasing dimensions of an object, leading to surface effects, such as surface tension, capillary forces, adhesion, and viscous drag.

Soil is a highly heterogeneous medium, consisting of a mixture of solid material, and of water- or air-filled pores [4] (Figure 1). Thus, soil can be theoretically interpreted, either as the (organized) arrangement of aggregates/particles in the soil (i.e., soil structure) [5,6], or as the connectivity, tortuosity, and heterogeneity of the pore space between these soil components (i.e., soil architecture) [7] (for a complete argumentation on soil structure versus soil architecture, see Baveye et al. [3]). Albeit the hierarchical organization of aggregates highly depends on the amount of energy that is applied to take the soil apart, soils can be classified based on the distribution of their different aggregate species (i.e., soil texture). The aggregates are generally divided depending on their size, i.e., from (very) stable micro-aggregates (<2 μm), which are mainly composed of organic matter and clays, to less stable macro-aggregates, which are commonly composed of silt (2–63 μm) or sand (63 μm to 2 mm) [8]. Physical and chemical processes driven by both biotic and abiotic factors cause changes in aggregate size and continuous particle rearrangement. These constant changes in spatial organization of the solids and voids affect the architecture and connectivity of the pore space, which, in turn, affects the distribution of water and gases [9], as well as the diffusion of substrates (e.g., organic matter) [10,11] and solutes (e.g., elements and ions) [12] in soil.

Water is one of the most important, but at the same time is most variable component of soil. Soil's water content depends on soil composition, rainfall, drainage, evaporation, temperature, and plant cover. Water is the medium connecting spatially separated areas in the soil matrix and becomes the solvent in which organic matter, microelements, and metabolites of different biological origins are dissolved or suspended (i.e., soil solution). The degree of water retention in soil microenvironments mainly depends on the soil pore neck size [13]. In macro-pores, localized in and between macro-aggregates, water is often well drained, whereas it is fully retained in micro-pores (localized between micro-aggregates) due to capillary action [13,14]. As a consequence, when alternating saturation/desiccation cycles that occur due to changes in precipitation or temperature, water status is more conserved in the micro-pores, generating fine water pockets rich in moisture that remain spatially disconnected from one another [14–18].

Micro-pores are also important for the retention of important biological molecules, such as organic matter, proteins, and nucleic acids [19–23]. As a result of the clay-cation exchange capacity (CEC) [4], they can all be adsorbed and retained by the negative charges of clay in soil micro-aggregates. Indeed, Ranjard and Richaume [14] found that organic matter is not homogenously distributed in soils, and higher concentrations (50–80%) were detected in micro-pores.

Furthermore, at the micro-scale level, micro-elements are to be considered for the growth and survival of microorganisms. Micro-elements are often present in the soil solution, as they are originating from both metabolically catalysed redox reactions as well as from abiotic chemical weathering of rock surfaces. They can diffuse in and out of the smallest pores, including the very narrow 1.8-nm-wide spaces between clay particles [3]. These diffusion processes cause pH and element concentration gradients that are highly dependent on the abiotic properties of the soil (e.g., the mineral composition, morphology, and texture), the geochemistry of the surrounding fluids, as well as the activity of microorganisms secreting highly reactive organic acids (e.g., oxalic acids, citric acid) [24].

Air (gases) resides in between the fractions of the soil pore network that are filled with the soil solution. The soil atmosphere depends on the connectivity of the non-water-filled pores of soil with the atmosphere or with other open pores. It usually consists of varying amounts of gasses such as oxygen, carbon dioxide, nitrogen, and nitrogen dioxide, but also of volatile organic compounds (gasses with biotic origin). The composition of the soil atmosphere highly depends on the production or consumption of a specific gas by local organisms, the solubility and diffusion of the gases in the soil water and, consequently, also on the capacity for water retention of the soil itself, as described before [13,25]. As an example, molecular oxygen has a very low solubility in water and its diffusion rate in water is 10,000-fold lower than in air [25]. Hence, soil micropores filled with water will rapidly turn anoxic upon consumption, limiting the growth and survival of many microbes. On the other hand, micropores can also offer protective microhabitats against "toxic" gases [26].

Independent of the soil component considered, the soil solution present in colloids can eventually diffuse out into wider pores, where it is readily available to microorganisms or is transported with the percolating water. It is also worth mentioning that filamentous fungi and plant roots stabilize the micro-aggregates of soil, and, in turn, together with other organisms living in the soil (e.g., nematodes, worms, larger animals), continuously increase soil solution conductivity through the soil. Nonetheless, the majority of substrates remain isolated and persist in the environment, possibly due to either physical protection or the separation from relevant enzymes [19]. In this regard, given their even smaller size, exoenzymes released by bacteria, archaea, and fungi into the soil solution can diffuse in and out of tiny pores and have significant roles both in the total fitness of the microbial population as well as in geo-chemical nutrient cycles of an ecosystem [27].

Respiration and decomposition rates differ considerably between sandy soils and soils rich in clay. This illustrates impressively how the dynamic interaction between the physical, air, and water components of soil directly influence the access, and thus the activity, of microbial populations to their substrate both in space and time. Sandy soils have lower surface-to-volume ratios than clay soils, meaning less water retention, but also lower segregation between microenvironments. In this context, soil heterogeneity has to be considered. For example, organic matter placed in different regions of the pore network was shown to be decomposed at different rates [28] while a number of other studies reported increased respiration rates only after disruption of the soil structure in soils with high clay content (e.g., [29]). These results taken together suggest that both the local environmental conditions as well as the uneven distribution of microbial populations within the local environment strongly affect the mineralization rates [28,30,31] and the range over which organisms can disperse and interact.

### 2.2. *Bacterial–Fungal Interactions: A Harsh Existence*

At the cell-to-cell level, bacteria and fungi interact at many different levels of intimacy that can be considered from two perspectives: in terms of physical associations and in terms of molecular communication [32]. Physical associations can range from seemingly disordered polymicrobial communities (i.e., biofilms) to highly specific symbiotic associations of fungal hyphae and bacterial cells (i.e., ectosymbiotic or endosymbiotic). Molecular interactions, on the other hand, involve a complex and diverse set of chemicals and compounds, and the interaction can be contextualized as antibiosis, signaling, and chemotaxis, metabolite exchange, metabolic conversion, adhesion, protein secretion, genetic exchange, and physicochemical changes. More often than not, multiple mechanism of interaction can be employed by one microorganism. The multitude of interactions and their effect on the partners or surrounding environment is extensive and well reviewed in [1,32–34].

Irrespective of the type of interaction, bacteria and fungi need to recognize each other prior to initiating any kind of target-oriented interaction. However, sensing and recognition mechanisms are often hindered by the particularity of soil and by the phenotypical characteristics of the organism (Figure 1).

When considering bacteria, they have their highest diversity in soil micro-aggregates, where they are either "swimming" in the soil solution, or attached to soil particles. This is not surprising, as it is in accordance with the higher concentrations of organic matter found in the micro-pores, the more stable conditions in water content, and a lower predation pressure by protozoan or other predators [35–39]. One of the first studies focusing on the distribution of bacteria in soil revealed that specific bacterial populations are typically residing inside micro-aggregates, with 40–70% of these bacteria being localized in the 2–20 μm and in <2 μm size micro-pores [14]. In some cases, cells can even penetrate pores smaller than themselves [40]. In contrast, more recent studies reported that pores in the 30 to 150 μm size range harbor a greater abundance of specific bacterial groups [41]. However, studies are scarce and only represent a snapshot in time, in a very dynamic environment. In fact, microbes explore a constantly changing environment in search of their *realized niche*, let it be by either active exploration of soil by fungal hyphae or by chemotactic movement, or passive transportation in bacteria. Any potential in bacterial mobility is limited by surface tension, capillary forces, and viscous drag that increase the energy requirement for their motility, particularly in partially saturated pore networks. Motility was found to cease, virtually completely, when the thickness of the water film was smaller than 1.5 μm [42]. For these reasons, microbial cells are often found as individual cells when associated with highly localized dissolved organic matter, or as patches of dense populations when linked to more conspicuous substrates [43]. In bulk soil, the average distance between neighbouring bacterial cells was found to be around 12.46 μm, with inter-cell distances shorter near the soil surface (10.38 μm) than at depth (>18 μm), due to changes in cell densities [44]. Simplified calculations suggests that, despite the very high number of cells and species in soil ($10^8$ cells per gram of soil), the number of neighbours that a single bacterial cell has within an interaction distance of ca. 20 μm is relatively limited (120 cells on average) [44]. Similar crude estimates were also found when the surfaces of soil pores were used to calculate the exclusion zone of cells on soil aggregates (a radius of 178 μm) [45].

A completely different story is depicted when filamentous fungi are considered. Fungi actively explore the soil pore space through hyphal spread, and cope well with heterogeneous distribution of nutrients [46]. They can cross air–water interfaces and nutrient-depleted spots to gain access to distant nutrient resources. The extensive mycelial architecture enables fungi to easily and efficiently re-allocate useful compounds to substrate-limited regions, to the benefit of exploratory colonisation of more unfavourable habitats [47].

A relevant feature of microbial interaction related to mycelial growth is that hyphae serve as dispersal vectors for motile bacteria (i.e., fungal highways) [48] and, hence, allow for bacterial colonization of new micro-habitats [49,50]. Moreover, the mycelia of fungi and oomycetes enhance bacterial activity by nutrient and water transfer (excretion) from the hyphae to the bacterial cells, thus enabling bacterial growth in otherwise too oligotrophic habitats [51], or enhancing microbial activity in dry soils [52]. An important example of such a symbiotic interaction is extraradical mycelia of arbuscular mycorrhizal and other mycorrhizal fungi in the rhizosphere of plants, where a large array of bacteria can directly consume fungal exudates released into the environment [34]. The mutualistic relationship can also be reciprocal: fungi take advantage of their bacterial partners to improve their carbon source pool in a mechanism named bacterial farming [53]. Fungal hyphae can also become an ideal hotspot for horizontal gene transfer [54–56] or bacterial prey populations [57], as they facilitate dispersal and preferential contact of bacteria in the hyphosphere. In this regard, it should be mentioned that, apart from fungal hyphae, plant roots and dead organisms also behave as hot spots for highly interactive (e.g., competition, pathogeny, mutualism, predation) microbial communities. Nonetheless, these many physical interactions between bacteria and hyphae-forming fungi may represent short-lived associations, as microscale communities frequently assemble and disassemble by migration, attachment, and detachment from surfaces and cells.

In this quite lonesome and unforgiving scenario, physical contact between cells is often not realizable. Hence, molecular compounds are employed and secreted in the soil solution to sense potential partners in their surroundings. Importantly, the release of molecular signals represent a cost, both energetic and elemental, and hence are regulated to maximize the fitness of the organism [27]. As molecules diffuse freely from any pore and are easily removed from the microenvironments, molecular interactions can only occur at relatively short distances. On leaf surfaces, for example, interactions among bacteria have been found to occur principally in the 5 to 20 μm range [58], whereas in soil, where local patches of cells may reside in a pocket of soil solution, diffusible metabolites can reach neighbouring cells up to 100 μm away [44,59]. The interactive dynamics are very different in soil crusts, biofilms, or mats where cells are well physically constrained and in direct contact with each other (e.g., [60]).

Sensing and recognition of fungi and bacteria includes also solution-independent ways. Volatile organic compounds (VOCs) are one important quorum sending vehicle-enabling communication over longer distances, especially in vision of the spatial separation between soil particles often occurring in unsaturated water conditions [61,62].

Finally, it is worth mentioning that bacteria and fungi can also indirectly interact by modifying their microenvironment in ways that can positively or negatively affect their partners (i.e., niche modulation) [34]. For example, the acidification or the neutralization of pH values influences the solubility of soil nutrients (e.g., phosphorus [24]), thus inhibiting or stimulating overall bacterial growth and metabolism [63–65]. Nutrient depletion is one example of the many ways of indirect interaction. Iron depletion by the excretion of bacterial or fungal siderophores can negatively affect the performance of surrounding organisms, including plants [66,67].

In terms of BFI in different scenarios of soil heterogeneity, there is a limited number of studies available. They show both a random distribution of soil microorganisms and a high degree of microbial networking. We will now discuss methodologies and techniques well-suitable for observing microbial interactions and we hope to stimulate the reader to further expand the research in this field.

## 3. Get Your Hands Dirty: Methods for Studying BFI

Advances in molecular methods and bioinformatics have provided enormous benefits in enlightening BFI from many different perspectives. The availability of genomic methods, such as DNA sequencing, especially with high-throughput sequencing methods, enabled deeper insights into the soil microbial diversity, and contributed enormously to a more comprehensive database build-up, and thus to better network analysis inferences. They surely provide the theoretical background of what to expect from a specific soil sample and undoubtedly increased the knowledge on unculturable microorganisms (some of the relevant literature on the topic include work by [68–73]). However, virtually all -omics approaches entirely ignore the geometry of the pore space in soils or the characteristics of microenvironments (e.g., [26,74–76]). Even highly confident co-occurrence network analyses cannot distinguish between true ecological interactions and other non-random processes (e.g., cross-feeding versus niche overlap) [69]. Although benchmarking studies are few and urgently needed, their relatively high false-positive interaction rate can probably be decreased by omitting interactions that could physically not have occurred. Thus, our experimental approach should be adapted to make it a better fit for exploring the complexity of BFI in soil (Figure 2).

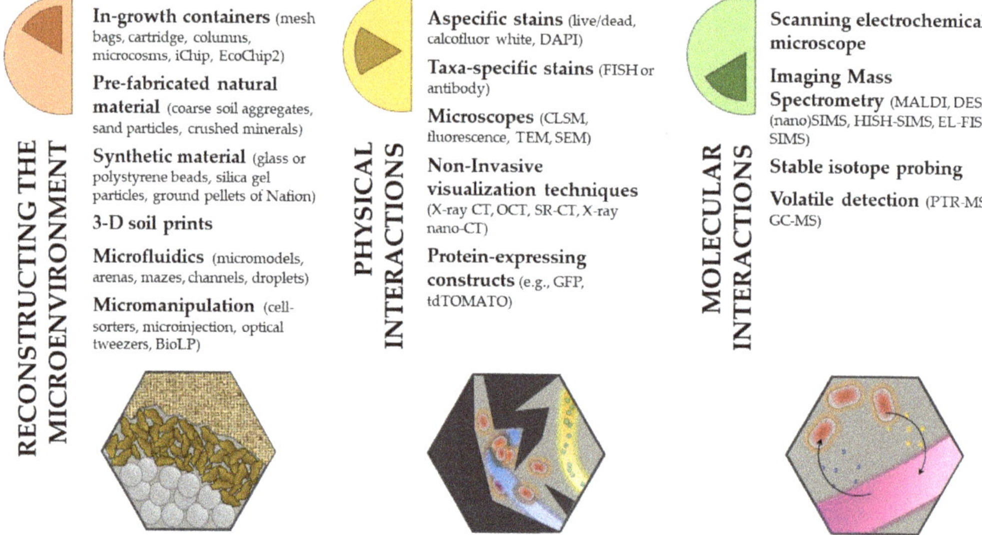


**Figure 2.** Overview of methods and technologies discussed in the review. They are divided by the applicability and information they can provide in regards to studies researching bacterial—fungal interactions. Often, by combining more than one together (e.g., microfluidics and FISH-SIP-nanoSIMS), information at different scales and levels are obtained.

As microorganisms interact at micro-scale, micro-scale data from microbiological, physical (e.g., porosity, water, air) and (bio)chemical (soil organic matter, micro-elements, pH) methods, including their new technological advances, should be combined to study BFI. As such, some of the methodologies discussed in the following chapters have also been applied to the investigation of abiotic and biotic aspects of the soil matrix [3], but here they will only be presented in vision of bacterial and fungal interactions. An overview of the abbreviations used to shorten the techniques' names is reported in Table 1.

**Table 1.** Abbreviations of technique's names discussed in the review.

| | | | |
|---|---|---|---|
| BIB-SEM | Broad ion-beam | GFP | Green fluorescent protein |
| BioLP | Biological laser printing | HISH | Halogen *in situ* hybridization |
| CARD-FISH | Catalysed reporter deposition | IMS | Imaging mass spectrometry |
| CLSM | Confocal laser scanning microscope | MALDI-IMS | Matrix-assisted laser desorption-ionization |
| DESI-IMS | Desorption electrospray ionization | OCT | Optical coherence tomography |
| DFA | Direct fluorescent antibody | PTR-MS | Proton transfer reaction |
| EL-FISH | Elemental labelling | SECM | Scanning electrochemical microscope |
| FACS | Fluorescence-activated cell sorter | SEM | Scanning electron microscope |
| FADS | Fluorescence-activated droplet sorter | (nano)SIMS | Secondary ion mass spectrometry |
| FIB-SEM | Focused ion-beam | SIP | Stable isotope probing |
| FISH | Fluorescence *in situ* hybridization | SR-CT | Synchroton computer tomography |
| FPT | Fluorescence probe techniques | TEM | Transmission electron microscope |
| GC-MS | Gas chromatography | X-ray CT | X-ray computer tomography |

Although not discussed, the combination of -omics techniques (e.g., metagenomics, transcriptomics, proteomics) alongside other standard protocols remains a very valuable strategy, provided that high-quality raw material is available. Furthermore, the reader should bear in mind other important methodological factors. It is essential to consider the heterogeneity of the soil for sampling strategies [26], the consistency of the experiment for reproducible results, and the applicability of the techniques for eventual high-throughput applications [77].

### 3.1. Reconstructing the Spatial Heterogeneity of Soil

In an ideal word, one would monitor in real time the dynamics of microbial interactions at the soil microscale, retrieving information about microbial taxonomy, distribution, and behaviour as well as about dependencies of these traits from the physicochemical properties of their micro-niches, directly in their natural microenvironment. Needless to say, this is far from the modern reality as it has proven impossible to quantify and analyse the interaction of microorganisms *in actual soil*.

Gause's co-culture experiments [78] are today a standard procedure for experimentally investigating microbial interactions. With inventive approaches (e.g., [79–85]), agar-based co-culture experiments re-create simplified communities in a controlled environment and thus provide ideal conditions to test ecological concepts concerning community stability and dynamics. However, a clear disadvantage of classical solid nutrient medium is the absence of habitat heterogeneity.

Artificial matrixes have been created that try to simulate as much spatial complexity in which BFI exists as possible but, at the same time, ease downstream analysis. Such soil matrices can often be sterilized, can be buried in soil within in-growth containers under natural conditions for a defined period of time, and then can be retrieved and transported to the laboratory with minimal disturbance. The applied in-growth method enables in situ cultivation of the local microbial communities [86,87] with the assumption that their interaction strategies are retained. Different designs of in-growth containers can be distinguished, such as nylon mesh bags with a mesh size >50 µm [88], or cartridge [87]. Depending on the downstream analysis to follow, they can be filled with different types of pre-fabricated natural material, such as macro-aggregates of soil fragments (5 mm-sized) [41], coarse sand particles [88,89], or mixtures of different crushed minerals [87]. These materials offer good field-representative soil aggregates that can be used also for in vitro experiments (e.g., [90]). However, they have the disadvantage of being opaque, such as real soil. This limits the range of applications to which they can be implemented, and thus are often substituted with artificial materials, as it will be described further down below.

Simon and colleagues [50], and more recently Junier and colleagues [91], developed the fungal highway column in order to detect bacterial dispersal by fungal hyphae in a forest soil. The fungal highway column consists of a small tube in which one medium section is in contact with the soil, while the second one, which is physically separated from the first, can only be colonized by bacteria when they are transported into it via fungal in-growth. Other sophisticated methods for in situ cultivation are iChip [92] and EcoChip2 [93]. They enable parallel cultivation and isolation of up to now uncultivated microbial species, and can elucidate the dependency of some organisms to one another, due to in situ growth factors, that cannot be simulated under standard laboratory conditions.

A common concern for all these types of in situ cultivation is the flow of water. Fungal hyphae are well-stabilized in the soil matrix, but bacteria can easily be transported or washed in and out of the in-growth containers by percolating water. This might ultimately bias BFI analyses. On the other hand, it has been shown that water often follows preferential pre-existing flow lines in the soil [94]. In addition, bacteria can tightly adhere to exogenous material, form biofilms, or be safely sheltered in micro-pores [26]. Ultimately, the real extent to which water can affect the BFI networks is still unknown.

A number of new manufacturing approaches have been developed to simulate the soil architecture, but with the advantages of reproducibility, accessibility, and laboratory manipulation. They are usually inoculated with defined fungal and bacteria isolates.

As it was aforementioned, synthetic materials, such as clean, spherical 500-µm (or bigger) glass beads [95–98]; crushed silica gel particles (irregularly shaped grains) [99,100]; polystyrene beads [101]; or ground pellets of Nafion [102–105], have been developed and widely applied as artificial porous media [106] to avoid the limitations given by the opacity of natural materials. Although soil is tremendously more heterogeneous (smaller pore size) and chemically complex, their low reactivity and their optical refractive index, similar to water, allows for the combination of co-cultivation with different visualization techniques (e.g., brightfield and fluorescence imaging, low-field magnetic resonance system, and X-ray CT), as discussed in the next section. Another interesting method for simulating soil microstructure is 3-D printing. Soil-like structures of nylon 12 or resin with parafilm wax [107,108] can be used to study the exploration and interaction strategies of different microbial inoculates, and may give future insight on BFI in these microhabitats. Some general precautions and limitations should be considered when setting up a microcosm experiment with synthetic soil manufacturing. The material chosen should be: biocompatible or at least not toxic to the organisms of interests, congruent with downstream imaging techniques, and an appropriate proxy for the environment of interest [3].

Microfluidic arrays can be deployed when more "personalized" structural constrains and dynamical chemical gradients are needed. The literature on this topic is extensive and more detailed information can be found in several reviews (e.g., [109–113]). Microfluidic platforms allow for the precisely organization and monitoring of small heterogeneous microbial populations [114,115] in a three-dimensional geometry [116]. They can be constructed in different dimensions (in volumes as small as ~100 fl [117]), materials (e.g., transparent polydimethylsiloxane (PDMS) [116,118,119], hydrogels, proteins crosslinked by multiphoton lithography [120], lipid-silica containers [121,122]), or shapes (e.g., soil micromodels [123,124], arenas [125], channels [126], mazes [127], or single droplets [128]). Some of the microfabricated biomaterials used for constructing a microfluidic system can be responsive to external stimuli, hence acting as both physical barriers, as well as an additional function for active control, manipulation, and observation of the microbes in real time. Some examples of dynamic parameters that can be tuned are pH, temperature, osmolarity, light intensity, as well as the geometry and the size of micro-chambers. Furthermore, the permeability of the walls and the rate of mass transport through the system can be manipulated in order to allow controlled diffusion of nutrients, waste products, or other small molecules [120,129]. The experiment can also be run at extensive temporal scales (e.g., miniaturized chemostats [130]), allowing continuous monitoring of the microbial dynamics over time. A series of limitations often highlighted by the authors comprise the sophisticated fabrication processes and complex technical set-up involved [110]. Furthermore, the porosity and complexity obtained with microfluidics are, as to be expected, lower than in real soil, hence it is necessary to compare the created environmental conditions to adequate references [111]. The majority of microfluidic devices have been used to study interactions in model organisms mainly between different bacterial cells (e.g., [117,131]), bacteria and plant [132], or fungi and other organisms [128,133,134]. A number of studies investigating the cell-to-cell interactions between fungi and bacteria are shyly emerging [135,136]. In these exclusive examples, interaction-induced physiological changes and metabolic exchange between fungal hyphae and their associated bacteria were investigated. Clearly, many more possibilities are awaiting.

Micromanipulation of single cells in time and space is also possible [117] that allows either their precise arrangement in a three-dimensional manner or their extraction from a complex heterogenous space. The precise arrangement of single cells can be achieved, for example, by fluorescence-activated droplet- or cell-sorter (FADS, FACS, respectively). These systems are based on the presence or absence of a fluorescent compound (e.g., GFP [137]), resorufin from resazurin [138], FISH-labelling, or stained antibodies [139], and work very well along with the set-up of microfluidic experiment where a defined combination of organisms distributed in specific locations can be achieved. Alternately, Partida-Martinez et al. [140] successfully implemented a microinjection technique based on a laser beam to re-introduce an endohyphal bacterium in its original curated fungal host and to later study their physiological states. Otherwise, to extract specific microorganisms from their microenvironment, micromanipulators of different types have been used [141–144], including the so-called 'optical tweezers' (e.g., [145,146]) and biological laser printing (BioLP) [147].

### 3.2. *Playing Hide and Seek with Bacteria and Fungi: Unravel Their Physical Interactions*

In mixed culture of bacteria and fungi growing in a matrix with defined topographical and chemical features providing the microorganisms with the opportunity to spatially organize, a set of methodologies can be applied to visualize, identify, and monitor their interaction.

Basic techniques for visualization of selected microbes in a matrix include non-specific stains, such as live-dead staining [148], stains with calcofluor white M2R [149,150], or DAPI [151]. These allow to spatially visualize and quantify bacteria and fungi in close proximity with minimal effort [44], although taxonomical information is not provided.

Hence, more advanced fluorescence-conjugated techniques are permanently being developed. They can be based on fluorescent probes targeting distinct short nucleotide sequences (fluorescence probe technique (FPT)), or on direct fluorescent antibodies targeting a specific protein domain (direct fluorescent antibody (DFA)) [139]. Fluorescence in situ hybridization (FISH) is a type of FPT that combines the phylogenetic identification of the taxon targeted on a single cell basis (e.g., prokaryotic cells) with the visualization of their distribution in situ (e.g., [152–154]). The technique is a pillar in molecular biology, and countless applications and specific developments exists (see for example the reviewing book by Azevedo and colleagues just published [155]). As it will be described in the next section, FISH as well as DFA are widely used in combination with many classical or modern applications in the recognition of taxa involved in fungal and bacterial molecular interplay. When dealing with a very bright background of fluorescing soil constituents, FISH probing can be combined with tyramide signal amplification. This technique is called catalysed reporter deposition (CARD)-FISH, and is very useful for obtaining higher signal intensities and reduced background interference [156–158]. The combination of CARD-FISH with confocal laser scanning microscopes (CLSM) allows quantification, localization, and visualization of microbial cells with even more depth-resolution. Transmission and scanning electron microscopes (TEM and SEM, respectively) can also be combined with cellular dyes in order to obtain higher resolutions for both qualitative and quantitative assessment of the targeted organisms. For example, a CARD-based approach using gold nanoparticles (GOLD-FISH) can be used when dealing with very small portions of soil [159].

Fixation techniques for whole soil blocks with special chemicals can be used in order to study the distribution of microorganisms in their specific microenvironments. Blocks of soils are first impregnated with resin. Very thin layers of material can then be serially removed either by a traditional microtome, or by broad or focused ion-beam scanning electron microscopy (BIB- or FIB-SEM) [150]. Alternately, based on the same principle, thin sections of fixed soil can be cut with a diamond-tipped blade and mounted on a microscope slide. If microbes have been previously stained, the soil organisms of interest can be specifically observed [160]. By merging the different layers, three-dimensional images can then be reconstructed from a series of z-dependent two-dimension images with special software and statistical interpolation techniques [160–162]. Stains and dyes are not without limitations, however. They often unspecifically bind to soil components (including organic matter), thus increasing the background noise that, together with auto-fluorescent objects naturally occurring in the sample, make the recognition of cells in a soil matrix challenging. This is especially true for bacterial and archaeal cells in respect to fungal hyphae, whose physical arrangement and spatial distribution are relatively more straightforward to identify. In addition, bleaching caused during image acquisition and the impossibility to view "the whole picture" in a single ocular field can be an additional problem. Visualization techniques do generally require long processing times, a lot of handling skills, and result interpretation experience. This probably explains why there is still a relatively low number of articles published on BFI, especially given that these tools have been around for already few decades. Lastly, but most importantly, a fundamental problem remains: when samples are physically altered and chemically fixed (i.e., killing), it is impossible to monitor microbial cell dynamics in situ and in vivo.

With regard to non-invasively studies on soil microbes in their undisturbed, structured environment, several new technical approaches have been developed and proposed during recent years: X-ray computed tomography (CT) [3,111] and its variations (e.g., optical coherence tomography (OCT) [60], synchroton tomography (SR-CT) [163], and X-ray nano-tomography [164]) have been used, in combination with other methods, to study the reciprocal interaction between microbes and their surrounding environment, in both native and artificial soils [41,96–98,102,165]. With X-ray CT, for example, the effect of the soil environment on fungal exploration behaviour and gas release was investigated by repeated scanning of microcosm systems over several weeks [90]. This type of approach was also adopted for direct observation of plant–pathogen interactions [166], and to detect the migratory capabilities of *Pseudomonas fluorescens* [167,168] in situ at high spatial resolution for the first time. However, studies on BFI implementing these techniques have lacked knowledge up to now. A fundamental problem of radiation-based imaging is that it is ionising and penetrative, meaning that it could possibly damage both structures and microorganisms within the substrate [164]. On the other hand, low irradiation doses applied to plant–soil interfaces were reported to neither influence root growth [169], nor microbial cell number, microbial community structure, or their potential activity [165,170]. Nevertheless, even if not lethal, care should be taken for the insurgence of mutations with unpredictable consequences. Lastly, the imaging resolution chosen can limit either the range or the feature of detection [111].

When concerned with the in vivo study of dynamically interacting organisms, simple and reliable applications for the live-cell screening of microbial organisms is the use of fluorescent protein-expressing bacteria (e.g., GFP, [85,142,171]) or fungi (e.g., tdTomato, [172]). A combination of red and green fluorescent proteins allows for a simultaneous detection of multiple organisms, and can be used to follow BFI both in situ and in vivo [172,173]. Although these modern molecular tools only permits to work with isolates or model organisms, if combined with a relevant experimental set-up, such as microfluidics, video microscopy, or other techniques, described in the next section, they can give a great deal of advantages [174].

### 3.3. *What Are They Doing? Investigating the Molecular Interactions in BFI*

The ability to confine small numbers of microorganisms and determine their location in space is an important step towards understanding the impact of spatial structure on microbial behaviour. However, the development of techniques for the in situ measurement of small molecule signals, and for the detection of other factors responsible for behavioural modulation, are equally important. These techniques are applicable to simple bacterial and fungal communities and ultimately allow cell-to-cell resolution.

Scanning electrochemical microscopy (SECM) can be considered as the most straightforward technique to quantify and spatially map the concentration of specific redox-active molecules (i.e., one compound at the time) proximal to populations of cells [175], such as the small molecules that are involved in the interaction in multispecies biofilms [176].

For more comprehensive studies, with imaging mass spectrometry (IMS), in all its variations, one can detect and describe multiple molecules at the same time, and superimpose their spatial distribution onto optical or fluorescence images of the sample. To put these advantages in perspective, a key difference between IMS and standard fluorescence microscopy is the capability of IMS to detect up to thousands of unique signals from one biological substrate, compared to only up to eight in the average fluorescence microscopes [177]. The interplay between competing organisms, the metabolic exchange in complex intra- and inter- species signaling, or the detection of new and uncharacterized molecules can be studied with this IMS technique. Furthermore, the direct link between the distribution of detected compounds and the observed phenotypes of the biological sample may give essential information on the function of the molecules themselves. Finally, the endogenous molecules can directly be analysed in their microenvironment, without the indirect biases that artificial substrates may cause (e.g., isotope feeding experiments) [177–179].

Matrix-assisted laser desorption-ionization (MALDI) [180] and desorption electrospray ionization (DESI) [181] are two types of IMS techniques that suit particularly well for studying the production of natural products by microorganisms in the laboratory [178]. Whereas the soft and absorbing nature of agar medium limits DESI-IMS analysis, with MALDI-IMS microorganisms can be grown on a 0.5–1.5-mm layer of agar for a defined period of time before being covered with a matrix and subjected to the analyses. Albeit certain media that have a high salt or sugar content may prove difficult to analyse, owing to ion suppression or uneven matrix crystallization, the use of agar-based IMS approach can provide important time-dependent correlations [180]. MALDI- and DESI-IMS have been widely applied for the characterization of microbial monocultures, such as cyanobacteria [182], bacterial colonies [183], or fungal–fungal interaction [184], and their application in BFI studies is slowly increasing, especially for the detection of antifungal compounds in co-culturing experiments [185–187]. Some of the limitations of these techniques derive from biases due to sample preparation, the simplification of otherwise complex environments, and signal distortion from background noise [179]. Secondary ion mass spectrometry (SIMS) [188] has the highest resolution (order of nanometres) of all the IMS techniques due to its extremely narrow ion beam. It can be applied to have either a focus on the analytical mass range (static) or to have a greater spatial resolution (dynamic). SIMS platforms can also perform very fine depth-profiling experiments by etching away the sample surface with the ion beam [178]. When an ultra-sensitive characterization of the molecular microenvironment between interacting cells is needed (down to 50 nm of resolution), nanoSIMS can be used. Obviously, with this level of resolution, the sampling area that can be analysed decreases to a few $mm^2$. Sample preparation is also a fundamental step. Samples should be mounted on a conductive surface and be extremely flat, although a notable exception was given by Vaidyanathan et al. [189].

Although experiments with these new modern techniques are on the rise, stable isotopes feed to a biological sample is still the most widespread technique to follow metabolic activities and interactions of bacterial–fungal communities in their natural environments. The literature available on stable isotope probing (SIP) is extensive and detailed descriptions can be found elsewhere (e.g., [141,190,191]). An incredible advantage of isotope labelling is that the substrate of interest can be directly fed in any artificial set-up chosen to simulate the (micro)-habitats of the organisms of interest (e.g., nature, microcosms, artificial matrixes). For example, different $^{13}C$-marked carbon sources were supplied in microcosms with soil material to monitor respiration rates and C utilization by bacteria and fungi [192]. Using airtight containers, Pinto-Tomás et al. [193] discovered that the nitrogen enriched material in the fungus garden of soil leaf-cutter ants was not derived by their extensive foraging activity, but was instead fixed by $N_2$-fixing bacteria hosted by the fungus itself. Another neat example that shows the intricate behaviour of fungi and bacteria is the study of Pion et al. [53] where, by tracing $^{13}C$-substrate, the researchers found out that the ascomycete fungus *Morchella crassipes* first farms *Pseudomonas putida* bacteria and then use them as carbon source. The precision and sensitivity of the instruments for the downstream analysis will eventually limit the experimental design. In fact, a common problematic intrinsic to all labelling experiments is the insufficient incorporation rate for detection. This can be avoided by long incorporation time, but in turn it increases the risk of cross-feeding i.e., contaminations [139]. Independently from the isotopically labelled substrates applied ($^{13}C$, $^{15}N$, $^{32}P$, $^{2}H$) and the technology used to detect the isotope incorporation (e.g., mass spectrometer, isotope ratio mass spectrometer, Raman microspectrometry, SIMS) [190], the monitoring of uptake and transfer of metabolites has to always be linked to the identification of the involved microorganisms. This can be achieved by a variety of different methods applied after isotope incorporation. Soil microorganisms can be discriminated based on taxa-specific biomolecules by phosphor lipid fatty acid analysis (PFLA); sterol analysis (e.g., using ergosterol as a biomarker for fungi) DNA-, RNA-, or protein-based methods; by FISH-microautoradiography [194,195]; or isotopic rRNA-arrays [196]. Molecular approaches have also been used for tracing back a specific taxon with labelled DNA, including PCR amplification as well as sequencing [197–199]. It is also possible to directly clone isotope-labelled DNA and then sequence it [200].

An exciting alternative to these classical taxa-discriminating approaches combines taxa-specific target probes (e.g., FISH, CARD-FISH, DFA) and stable isotope probing (SIP) with the simultaneous detection of atoms produced by nanoSIMS. If based on creative experimental designs, the combination of these techniques can substantially help for linking identity to functions by tracking the uptake and transfer of isotopically labelled compounds in environmental samples where the individual taxa cannot be isolated from one another. A satisfactory overview of these combination methods is given by Musat et al. [201]. Briefly, a microbial community that was grown with metabolic-labelled tracers (e.g., $^{13}C$ or $^{15}N$) is fixed, hybridized with phylogenetic probes, and visualized by fluorescence microscopy to obtain a picture of the distribution of specific taxa in space. Then, the sample is analysed by nanoSIMS [155] which maps the existing molecules in the sample by detecting and identifying them from their specific locations. Finally, a superimposed picture of the partners is obtained, which involved giving both physical and molecular information on the interaction.

The use of SIP-nanoSIMS has great potential as an attractive alternative to stand-alone autoradiography experiments for the study of the molecular micro-environments of microbes, also thanks to the increase in more accessible facilities and the decrease in costs. An increasing number of studies have used the combination of these powerful techniques to link identity and metabolic behaviours within bacterial communities (e.g., [202,203]), between microbe–host interactions (e.g., [204]) or in cyanobacteria mats [205,206]. Less studied are the interaction between bacteria and fungi. An exceptional example in this regard was carried out by Worrich et al. [51] where they investigated the water and nutrient exchange between bacterial cells and fungal hyphae in stress conditions.

Direct label imaging during nanoSIMS analysis is also emerging with the use of EL-FISH nanoSIMS [159] or HISH-SIMS [207], although the studies for microbial interactions using these techniques are still limited.

Finally, but not least importantly, it is worth mentioning that interesting insights in fungal and bacterial communication can be retrieved from instruments that detect VOC (proton transfer reaction (PTR)-MS or GC-MS) [208]. These molecules are very important when considering the porosity of the soil (or similar artificial models) and the connection that air provides among different locations, even between considerable distances.

## 4. Outlook

Although the list of tools provided in this review may seem extensive, the majority of the available studies carried out on bacterial–fungal interaction used a repetition of a few well-established, often simple, methods such as agar-plating, isotope probing, or fluorescence microscopy, combined in new neat ways to answer research questions never asked before. Surely, there were some exceptions, but negligible if compared to the long-time availability of novel methods such as CARD-FISH, SIP-nanoSIMS, or microfluidics, and the immense number of plausible combinations of these. All methods have strengths and weaknesses, as well as costs, thus the best combination of these applications remains the one that best fits the specific research questions that a researcher is trying to address. Nevertheless, to acquire the best realistic overview of the system, the combination of tools that aims at providing data at different levels of complexity is highly required (Figure 3). Inventiveness and perseverance are the way forward.

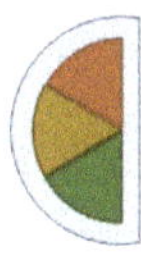

BACTERIAL – FUNGAL INTERACTIONS
(comprehensive research)

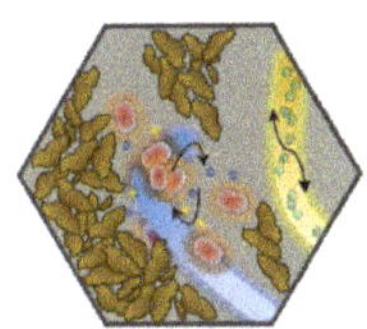

**Figure 3.** The more variables and data that an experimental design incorporates or collects at different levels of complexity (i.e., by reconstructing a coherent environmental set-up as well as by investigating the physical and the molecular interactions), the more comprehensive becomes our understanding of bacterial–fungal interactions, setting the field a step closer to accurately depict the reality of the natural environment.

Studies investigating the interaction of fungi and bacteria within each own taxonomical unit are greater than the studies investigating the combination of the two. This, along with the too often bypassed research that tries to untangle the complexity of the physicochemical properties of soils, shows that this field of microbiology often still works in sectorial ways. Just as much there *really* is a community of microbes living together for their "greater good" that cannot be extracted from their environmental background without seeing it biased, we, as observers, should learn from them and work together, in an interdisciplinary way, to better catch a glimpse of their intricate lives.

**Funding:** This study was financed by the Open Access Funding by the Austrian Science Fund (FWF) and by the Land Tirol (project Microbial Interactions in Snow-covered soil MICINSNOW-P31038).

**Institutional Review Board Statement:** Not applicable.

**Informed Consent Statement:** Not applicable.

**Data Availability Statement:** Data is contained within the article.

**Conflicts of Interest:** No conflicts of interests to be declared.

## References

1. Jansson, J.K.; Hofmockel, K.S. The soil microbiome—From metagenomics to metaphenomics. *Curr. Opin. Microbiol.* **2018**, *43*, 162–168. [CrossRef]
2. Baldrian, P. The known and the unknown in soil microbial ecology. *FEMS Microbiol. Ecol.* **2019**, *95*, 1–9. [CrossRef]
3. Baveye, P.C.; Otten, W.; Kravchenko, A.; Balseiro-Romero, M.; Beckers, É.; Chalhoub, M.; Darnault, C.; Eickhorst, T.; Garnier, P.; Hapca, S.; et al. Emergent properties of microbial activity in heterogeneous soil microenvironments: Different research approaches are slowly converging, yet major challenges remain. *Front. Microbiol.* **2018**, *9*, 1929. [CrossRef]

4. Young, I.M. Interactions and self-organization in the soil-microbe complex. *Science* **2004**, *304*, 1634–1637. [CrossRef]
5. Dexter, A. Advances in characterization of soil structure. *Soil Tillage Res.* **1988**, *11*, 199–238. [CrossRef]
6. Hillel, D. *Introduction to Environmental Soil Physics*; Elsevier: Amsterdam, The Netherlands, 2003.
7. Young, I.; Crawford, J.; Rappoldt, C. New methods and models for characterising structural heterogeneity of soil. *Soil Tillage Res.* **2001**, *61*, 33–45. [CrossRef]
8. Tisdall, J.M.; Oades, J.M. Organic matter and water-stable aggregates in soils. *Eur. J. Soil Sci.* **1982**, *33*, 141–163. [CrossRef]
9. Horn, R.; Smucker, A. Structure formation and its consequences for gas and water transport in unsaturated arable and forest soils. *Soil Tillage Res.* **2005**, *82*, 5–14. [CrossRef]
10. Mueller, C.W.; Koelbl, A.; Hoeschen, C.; Hillion, F.; Heister, K.; Herrmann, A.M.; Kögel-Knabner, I. Submicron scale imaging of soil organic matter dynamics using NanoSIMS—From single particles to intact aggregates. *Org. Geochem.* **2012**, *42*, 1476–1488. [CrossRef]
11. Rawlins, B.G.; Wragg, J.; Reinhard, C.; Atwood, R.C.; Houston, A.; Lark, R.M.; Rudolph, S. Three-dimensional soil organic matter distribution, accessibility and microbial respiration in macroaggregates using osmium staining and synchrotron X-ray computed tomography. *SOIL* **2016**, *2*, 659–671. [CrossRef]
12. Jasinska, E.; Wetzel, H.; Baumgartl, T.; Horn, R. Heterogeneity of physico-chemical properties in structured soils and its consequences. *Pedosphere* **2006**, *16*, 284–296. [CrossRef]
13. Schurgers, G.; Dörsch, P.; Bakken, L.; Leffelaar, P.; Haugen, L. Modelling soil anaerobiosis from water retention characteristics and soil respiration. *Soil Biol. Biochem.* **2006**, *38*, 2637–2644. [CrossRef]
14. Ranjard, L.; Richaume, A. Quantitative and qualitative microscale distribution of bacteria in soil. *Res. Microbiol.* **2001**, *152*, 707–716. [CrossRef]
15. Otten, W.; Gilligan, C.; Watts, C.; Dexter, A.; Hall, D. Continuity of air-filled pores and invasion thresholds for a soil-borne fungal plant pathogen, *Rhizoctonia solani*. *Soil Biol. Biochem.* **1999**, *31*, 1803–1810. [CrossRef]
16. Otten, W.; Gilligan, C.A. Soil structure and soil-borne diseases: Using epidemiological concepts to scale from fungal spread to plant epidemics. *Eur. J. Soil Sci.* **2006**, *57*, 26–37. [CrossRef]
17. Kaisermann, A.; Maron, P.; Beaumelle, L.; Lata, J. Fungal communities are more sensitive indicators to non-extreme soil moisture variations than bacterial communities. *Appl. Soil Ecol.* **2015**, *86*, 158–164. [CrossRef]
18. Baveye, P.C.; Ebaveye, J.; Egowdy, J. Soil "ecosystem" services and natural capital: Critical appraisal of research on uncertain ground. *Front. Environ. Sci.* **2016**, *4*, 1–49. [CrossRef]
19. Demanèche, S.; Jocteur-Monrozier, L.; Quiquampoix, H.; Simonet, P. Evaluation of biological and physical protection against nuclease degradation of clay-bound plasmid DNA. *Appl. Environ. Microbiol.* **2001**, *67*, 293–299. [CrossRef]
20. Pietramellara, G.; Ascher, J.; Borgogni, F.; Ceccherini, M.T.; Guerri, G.; Nannipieri, P. Extracellular DNA in soil and sediment: Fate and ecological relevance. *Biol. Fertil. Soils* **2009**, *45*, 219–235. [CrossRef]
21. Romanowski, G.; Lorenz, M.G.; Sayler, G.; Wackernagel, W. Persistence of free plasmid DNA in soil monitored by various methods, including a transformation assay. *Appl. Environ. Microbiol.* **1992**, *58*, 3012–3019. [CrossRef]
22. Stotzky, G. Influence of soil mineral colloids on metabolic processes, growth, adhesion, and ecology of microbes and viruses. *Geochem. Soil Radionucl.* **2015**, *17*, 305–428. [CrossRef]
23. Nielsen, K.M.; Calamai, L.; Pietramellara, G. Stabilization of extracellular DNA and proteins by transient binding to various soil components. *Soil Biol.* **2006**, *8*, 141–157. [CrossRef]
24. Konhauser, K. *Introduction to Geomicrobiology*, 1st ed.; John Wiley & Sons: Hoboken, NJ, USA; Blackwell Publishing: Hoboken, NJ, USA, 2007.
25. Rappoldt, C.; Crawford, J. The distribution of anoxic volume in a fractal model of soil. *Geoderma* **1999**, *88*, 329–347. [CrossRef]
26. Lombard, N.; Prestat, E.; van Elsas, J.D.; Simonet, P. Soil-specific limitations for access and analysis of soil microbial communities by metagenomics. *FEMS Microbiol. Ecol.* **2011**, *78*, 31–49. [CrossRef]
27. Nunan, N.; Schmidt, H.; Raynaud, X. The ecology of heterogeneity: Soil bacterial communities and C dynamics. *Philos. Trans. R. Soc. B Biol. Sci.* **2020**, *375*, 20190249. [CrossRef]
28. Strong, D.T.; Wever, H.D.E.; Merckx, R.; Recous, S. Spatial location of carbon decomposition in the soil pore system. *Eur. J. Soil Sci.* **2004**, *55*, 739–750. [CrossRef]
29. Salomé, C.; Nunan, N.; Pouteau, V.; Lerch, T.Z.; Chenu, C. Carbon dynamics in topsoil and in subsoil may be controlled by different regulatory mechanisms. *Glob. Change Biol.* **2010**, *16*, 416–426. [CrossRef]
30. Ruamps, L.S.; Nunan, N.; Pouteau, V.; Leloup, J.; Raynaud, X.; Roy, V.; Chenu, C. Regulation of soil organic C mineralisation at the pore scale. *FEMS Microbiol. Ecol.* **2013**, *86*, 26–35. [CrossRef]
31. Ruamps, L.S.; Nunan, N.; Chenu, C. Microbial biogeography at the soil pore scale. *Soil Biol. Biochem.* **2011**, *43*, 280–286. [CrossRef]
32. Frey-Klett, P.; Burlinson, P.; Deveau, A.; Barret, M.; Tarkka, M.; Sarniguet, A. Bacterial–fungal interactions: Hyphens between agricultural, clinical, environmental, and food microbiologists. *Microbiol. Mol. Biol. Rev.* **2011**, *75*, 583–609. [CrossRef]
33. Bonfante, P.; Anca, I.-A. Plants, mycorrhizal fungi, and bacteria: A network of interactions. *Annu. Rev. Microbiol.* **2009**, *63*, 363–383. [CrossRef] [PubMed]
34. Deveau, A.; Bonito, G.; Uehling, J.; Paoletti, M.; Becker, M.; Bindschedler, S.; Hacquard, S.; Hervé, V.; Labbé, J.; Lastovetsky, O.A.; et al. Bacterial–fungal interactions: Ecology, mechanisms and challenges. *FEMS Microbiol. Rev.* **2018**, *42*, 335–352. [CrossRef] [PubMed]

35. Clarholm, M. Protozoan grazing of bacteria in soil—Impact and importance. *Microb. Ecol.* **1981**, *7*, 343–350. [CrossRef] [PubMed]
36. Kuikman, P.; van Elsas, J.; Jansen, A.; Burgers, S.; van Veen, J. Population dynamics and activity of bacteria and protozoa in relation to their spatial distribution in soil. *Soil Biol. Biochem.* **1990**, *22*, 1063–1073. [CrossRef]
37. Rutherford, P.M.; Juma, N.G. Influence of texture on habitable pore space and bacterial-protozoan populations in soil. *Biol. Fertil. Soils* **1992**, *12*, 221–227. [CrossRef]
38. Wright, D.A.; Killham, K.; Glover, L.A.; Prosser, J.I. Role of pore size location in determining bacterial activity during predation by protozoa in soil. *Appl. Environ. Microbiol.* **1995**, *61*, 3537–3543. [CrossRef]
39. Wright, D.; Killham, K.; Glover, L.; Prosser, J. The effect of location in soil on protozoal grazing of a genetically modified bacterial inoculum. *Geoderma* **1993**, *56*, 633–640. [CrossRef]
40. Männik, J.; Driessen, R.; Galajda, P.; Keymer, J.E.; Dekker, C. Bacterial growth and motility in sub-micron constrictions. *Proc. Natl. Acad. Sci. USA* **2009**, *106*, 14861–14866. [CrossRef]
41. Kravchenko, A.N.; Negassa, W.C.; Guber, A.K.; Hildebrandt, B.; Marsh, T.; Rivers, M.L. Intra-aggregate pore structure influences phylogenetic composition of bacterial community in macroaggregates. *Soil Sci. Soc. Am. J.* **2014**, *78*, 1924–1939. [CrossRef]
42. Dechesne, A.; Wang, G.; Gulez, G.; Or, D.; Smets, B.F. Hydration-controlled bacterial motility and dispersal on surfaces. *Proc. Natl. Acad. Sci. USA* **2010**, *107*, 14369–14372. [CrossRef]
43. Vos, M.; Wolf, A.B.; Jennings, S.J.; Kowalchuk, G.A. Micro-scale determinants of bacterial diversity in soil. *FEMS Microbiol. Rev.* **2013**, *37*, 936–954. [CrossRef] [PubMed]
44. Raynaud, X.; Nunan, N. Spatial ecology of bacteria at the microscale in soil. *PLoS ONE* **2014**, *9*, e87217. [CrossRef]
45. Pennell, K.D. Specific surface area. In *Reference Module in Earth Systems and Environmental Sciences*; Elias, S.A., Ed.; Elsevier: Amsterdam, The Netherlands, 2016; pp. 1–8.
46. Boswell, G.P.; Jacobs, H.; Ritz, K.; Gadd, G.M.; Davidson, F. The development of fungal networks in complex environments. *Bull. Math. Biol.* **2007**, *69*, 605–634. [CrossRef] [PubMed]
47. Fricker, M.D.; Heaton, L.L.M.; Jones, N.S.; Boddy, L. The mycelium as a network. *Fungal Kingd.* **2017**, *5*, 335–367. [CrossRef]
48. Kohlmeier, S.; Smits, T.H.; Ford, R.M.; Keel, C.; Harms, H.; Wick, L.Y. Taking the fungal highway: Mobilization of pollutant-degrading bacteria by fungi. *Environ. Sci. Technol.* **2005**, *39*, 4640–4646. [CrossRef]
49. Warmink, J.A.; Nazir, R.; van Elsas, J.D. Universal and species-specific bacterial 'fungiphiles' in the mycospheres of different basidiomycetous fungi. *Environ. Microbiol.* **2009**, *11*, 300–312. [CrossRef]
50. Simon, A.; Hervé, V.; Al-Dourobi, A.; Verrecchia, E.; Junier, P. An in situ inventory of fungi and their associated migrating bacteria in forest soils using fungal highway columns. *FEMS Microbiol. Ecol.* **2016**, *93*, fiw217. [CrossRef]
51. Worrich, A.; Stryhanyuk, H.; Musat, N.; König, S.; Banitz, T.; Centler, F.; Frank, K.; Thullner, M.; Harms, H.; Richnow, H.-H.; et al. Mycelium-mediated transfer of water and nutrients stimulates bacterial activity in dry and oligotrophic environments. *Nat. Commun.* **2017**, *8*, 15472. [CrossRef]
52. Guhr, A.; Borken, W.; Spohn, M.; Matzner, E. Redistribution of soil water by a saprotrophic fungus enhances carbon mineralization. *Proc. Natl. Acad. Sci. USA* **2015**, *112*, 14647–14651. [CrossRef] [PubMed]
53. Pion, M.; Spangenberg, J.E.; Simon, A.; Bindschedler, S.; Flury, C.; Chatelain, A.; Bshary, R.; Job, D.; Junier, P. Bacterial farming by the fungus *Morchella crassipes*. *Proc. R. Soc. B Boil. Sci.* **2013**, *280*, 20132242. [CrossRef] [PubMed]
54. Haq, I.U.; Zhang, M.; Yang, P.; van Elsas, J.D. The interactions of bacteria with fungi in soil. *Adv. Appl. Microbiol.* **2014**, *89*, 185–215. [CrossRef]
55. Berthold, T.; Centler, F.; Hübschmann, T.; Remer, R.; Thullner, M.; Harms, H.; Wick, L.Y. Mycelia as a focal point for horizontal gene transfer among soil bacteria. *Sci. Rep.* **2016**, *6*, 36390. [CrossRef] [PubMed]
56. Nazir, R.; Mazurier, S.; Yang, P.; Lemanceau, P.; van Elsas, J.D. The ecological role of type three secretion systems in the interaction of bacteria with fungi in soil and related habitats is diverse and context-dependent. *Front. Microbiol.* **2017**, *8*, 38. [CrossRef] [PubMed]
57. Otto, S.; Bruni, E.P.; Harms, H.; Wick, L.Y. Catch me if you can: Dispersal and foraging of *Bdellovibrio bacteriovorus* 109J along mycelia. *ISME J.* **2016**, *11*, 386–393. [CrossRef] [PubMed]
58. Esser, D.S.; Leveau, J.H.; Meyer, K.M.; Wiegand, K. Spatial scales of interactions among bacteria and between bacteria and the leaf surface. *FEMS Microbiol. Ecol.* **2015**, *91*, 1–13. [CrossRef] [PubMed]
59. Gantner, S.; Schmid, M.; Dürr, C.; Schuhegger, R.; Steidle, A.; Hutzler, P.; Langebartels, C.; Eberl, L.; Hartmann, A.; Dazzo, F.B. In situ quantitation of the spatial scale of calling distances and population density-independent N-acylhomoserine lactone-mediated communication by rhizobacteria colonized on plant roots. *FEMS Microbiol. Ecol.* **2006**, *56*, 188–194. [CrossRef]
60. Cai, P.; Sun, X.; Wu, Y.; Gao, C.; Mortimer, M.; Holden, P.A.; Redmile-Gordon, M.; Huang, Q. Soil biofilms: Microbial interactions, challenges, and advanced techniques for ex-situ characterization. *Soil Ecol. Lett.* **2019**, *1*, 85–93. [CrossRef]
61. Schmidt, R.; Cordovez, V.; de Boer, W.; Raaijmakers, J.; Garbeva, P. Volatile affairs in microbial interactions. *ISME J.* **2015**, *9*, 2329–2335. [CrossRef]
62. Effmert, U.; Kalderás, J.; Warnke, R.; Piechulla, B. Volatile mediated interactions between bacteria and fungi in the soil. *J. Chem. Ecol.* **2012**, *38*, 665–703. [CrossRef]
63. Nazir, R.; Warmink, J.A.; Boersma, H.; van Elsas, J.D. Mechanisms that promote bacterial fitness in fungal-affected soil microhabitats. *FEMS Microbiol. Ecol.* **2009**, *71*, 169–185. [CrossRef]
64. Bignell, E. The molecular basis of pH sensing, signaling, and homeostasis in fungi. *Adv. Appl. Microbiol.* **2012**, *79*, 1–18. [PubMed]

65. Braunsdorf, C.; Mailänder-Sánchez, D.; Schaller, M. Fungal sensing of host environment. *Cell. Microbiol.* **2016**, *18*, 1188–1200. [CrossRef]
66. Lurthy, T.; Cantat, C.; Jeudy, C.; Declerck, P.; Gallardo, K.; Barraud, C.; Leroy, F.; Ourry, A.; Lemanceau, P.; Salon, C.; et al. Impact of bacterial siderophores on iron status and ionome in pea. *Front. Plant Sci.* **2020**, *11*, 1–12. [CrossRef] [PubMed]
67. Haas, H. Fungal siderophore metabolism with a focus on *Aspergillus fumigatus*. *Nat. Prod. Rep.* **2014**, *31*, 1266–1276. [CrossRef] [PubMed]
68. Barberán, A.; Bates, S.T.; Casamayor, E.O.; Fierer, N. Using network analysis to explore co-occurrence patterns in soil microbial communities. *ISME J.* **2011**, *6*, 343–351. [CrossRef] [PubMed]
69. Faust, K.; Raes, J. Microbial interactions: From networks to models. *Nat. Rev. Genet.* **2012**, *10*, 538–550. [CrossRef]
70. Jiang, D.; Armour, C.R.; Hu, C.; Mei, M.; Tian, C.; Sharpton, T.J.; Jiang, Y. Microbiome multi-omics network analysis: Statistical considerations, limitations, and opportunities. *Front. Genet.* **2019**, *10*, 995. [CrossRef]
71. Matchado, M.S.; Lauber, M.; Reitmeier, S.; Kacprowski, T.; Baumbach, J.; Haller, D.; List, M. Network analysis methods for studying microbial communities: A mini review. *Comput. Struct. Biotechnol. J.* **2021**, *19*, 2687–2698. [CrossRef]
72. Weiss, S.; van Treuren, W.; Lozupone, C.; Faust, K.; Friedman, J.; Deng, Y.; Xia, L.C.; Xu, Z.Z.; Ursell, L.; Alm, E.J.; et al. Correlation detection strategies in microbial data sets vary widely in sensitivity and precision. *ISME J.* **2016**, *10*, 1669–1681. [CrossRef]
73. Telagathoti, A.; Probst, M.; Peintner, U. Habitat, snow-cover and soil pH, affect the distribution and diversity of mortierellaceae species and their associations to bacteria. *Front. Microbiol.* **2021**, *12*, 669784. [CrossRef]
74. Nannipieri, P.; Ascher, J.; Ceccherini, M.T.; Landi, L.; Pietramellara, G.; Renella, G. Microbial diversity and soil functions. *Eur. J. Soil Sci.* **2003**, *54*, 655–670. [CrossRef]
75. Nannipieri, P.; Ascher, J.; Ceccherini, M.T.; Landi, L.; Pietramellara, G.; Renella, G. Landmark Papers: No. 6. *Eur. J. Soil Sci.* **2017**, *70*, 1. [CrossRef]
76. Nunan, N. The microbial habitat in soil: Scale, heterogeneity and functional consequences. *J. Plant Nutr. Soil Sci.* **2017**, *180*, 425–429. [CrossRef]
77. Cordero, O.X.; Datta, M.S. Microbial interactions and community assembly at microscales. *Curr. Opin. Microbiol.* **2016**, *31*, 227–234. [CrossRef]
78. Gause, G.F. *The Struggle for Existence*; Williams and Wilkins Co.: Baltimore, MD, USA, 1934.
79. Bravo, D.; Cailleau, G.; Bindschedler, S.; Simon, A.; Job, D.; Verrecchia, E.; Junier, P. Isolation of oxalotrophic bacteria able to disperse on fungal mycelium. *FEMS Microbiol. Lett.* **2013**, *348*, 157–166. [CrossRef] [PubMed]
80. Ingham, C.J.; Kalisman, O.; Finkelshtein, A.; Ben-Jacob, E. Mutually facilitated dispersal between the nonmotile fungus Aspergillus fumigatus and the swarming bacterium *Paenibacillus vortex*. *Proc. Natl. Acad. Sci. USA* **2011**, *108*, 19731–19736. [CrossRef] [PubMed]
81. Krug, L.; Erlacher, A.; Markut, K.; Berg, G.; Cernava, T. The microbiome of alpine snow algae shows a specific inter-kingdom connectivity and algae-bacteria interactions with supportive capacities. *ISME J.* **2020**, *14*, 2197–2210. [CrossRef] [PubMed]
82. Palmieri, D.; Vitale, S.; Lima, G.; di Pietro, A.; Turrà, D. A bacterial endophyte exploits chemotropism of a fungal pathogen for plant colonization. *Nat. Commun.* **2020**, *11*, 1–11. [CrossRef]
83. Schmidt, R.; de Jager, V.; Zühlke, D.; Wolff, C.; Bernhardt, J.; Cankar, K.; Beekwilder, J.; van Ijcken, W.; Sleutels, F.; de Boer, W.; et al. Fungal volatile compounds induce production of the secondary metabolite sodorifen in *Serratia plymuthica* PRI-2C. *Sci. Rep.* **2017**, *7*, 1–14. [CrossRef]
84. Stopnišek, N.; Zühlke, D.; Carlier, A.; Barberán, A.; Fierer, N.; Becher, D.; Riedel, K.; Eberl, L.; Weisskopf, L. Molecular mechanisms underlying the close association between soil Burkholderia and fungi. *ISME J.* **2015**, *10*, 253–264. [CrossRef]
85. Aspray, T.J.; Jones, E.E.; Davies, M.W.; Shipman, M.; Bending, G.D. Increased hyphal branching and growth of ectomycorrhizal fungus *Lactarius rufus* by the helper bacterium *Paenibacillus* sp. *Mycorrhiza* **2013**, *23*, 403–410. [CrossRef]
86. Dresch, P.; Falbesoner, J.; Ennemoser, C.; Hittorf, M.; Kuhnert, R.; Peintner, U. Emerging from the ice-fungal communities are diverse and dynamic in earliest soil developmental stages of a receding glacier. *Environ. Microbiol.* **2019**, *21*, 1864–1880. [CrossRef]
87. Casar, C.P.; Kruger, B.R.; Flynn, T.M.; Masterson, A.L.; Momper, L.M.; Osburn, M.R. Mineral-hosted biofilm communities in the continental deep subsurface, Deep Mine Microbial Observatory, SD, USA. *Geobiology* **2020**, *18*, 508–522. [CrossRef]
88. Wallander, H.; Nilsson, L.O.; Hagerberg, D.; Bååth, E. Estimation of the biomass and seasonal growth of external mycelium of ectomycorrhizal fungi in the field. *New Phytol.* **2001**, *151*, 753–760. [CrossRef]
89. Harvey, H.; Wildman, R.D.; Mooney, S.J.; Avery, S.V. Soil aggregates by design: Manufactured aggregates with defined microbial composition for interrogating microbial activities in soil microhabitats. *Soil Biol. Biochem.* **2020**, *148*, 107870. [CrossRef]
90. Helliwell, J.; Miller, T.; Whalley, R.; Mooney, S.; Sturrock, C. Quantifying the impact of microbes on soil structural development and behaviour in wet soils. *Soil Biol. Biochem.* **2014**, *74*, 138–147. [CrossRef]
91. Junier, P.; Cailleau, G.; Palmieri, I.; Vallotton, C.; Trautschold, O.C.; Junier, T.; Paul, C.; Bregnard, D.; Palmieri, F.; Estoppey, A.; et al. Democratization of fungal highway columns as a tool to investigate bacteria associated with soil fungi. *FEMS Microbiol. Ecol.* **2021**, *97*, fiab003. [CrossRef] [PubMed]
92. Nichols, D.; Cahoon, N.; Trakhtenberg, E.M.; Pham, L.; Mehta, A.; Belanger, A.; Kanigan, T.; Lewis, K.; Epstein, S.S. Use of ichip for high-throughput in situ cultivation of "uncultivable" microbial species. *Appl. Environ. Microbiol.* **2010**, *76*, 2445–2450. [CrossRef] [PubMed]

93. Das, P.S.; Gagnon-Turcotte, G.; Ouazaa, K.; Bouzid, K.; Hosseini, S.N.; Bharucha, E.; Tremblay, D.; Moineau, S.; Messaddeq, Y.; Corbeil, J.; et al. The EcoChip 2: An autonomous sensor platform for multimodal bio-environmental monitoring of the northern habitat. In Proceedings of the 42nd Annual International Conferences of the IEEE Engineering in Medicine and Biology Society, Montreal, QC, Canada, 20–24 July 2020; pp. 4101–4104. [CrossRef]
94. Zhang, Y.; Zhang, Z.; Ma, Z.; Chen, J.; Akbar, J.; Zhang, S.; Che, C.; Zhang, M.; Cerdà, A. A review of preferential water flow in soil science. *Can. J. Soil Sci.* **2018**, *98*, 604–618. [CrossRef]
95. Lilje, O.; Lilje, E.; Marano, A.V.; Gleason, F.H. Three dimensional quantification of biological samples using micro-computer aided tomography (microCT). *J. Microbiol. Methods* **2013**, *92*, 33–41. [CrossRef]
96. Davit, Y.; Iltis, G.; Debenest, G.; Veran-Tissoires, S.; Wildenschild, D.; Gerino, M.; Quintard, M. Imaging biofilm in porous media using X-ray computed microtomography. *J. Microsc.* **2010**, *242*, 15–25. [CrossRef] [PubMed]
97. Iltis, G.C.; Armstrong, R.T.; Jansik, D.P.; Wood, B.D.; Wildenschild, D. Imaging biofilm architecture within porous media using synchrotron-based X-ray computed microtomography. *Water Resour. Res.* **2011**, *47*, 1–5. [CrossRef]
98. Peszynska, M.; Trykozko, A.; Iltis, G.; Schlueter, S.; Wildenschild, D. Biofilm growth in porous media: Experiments, computational modeling at the porescale, and upscaling. *Adv. Water Resour.* **2016**, *95*, 288–301. [CrossRef]
99. Sanderlin, A.B.; Vogt, S.; Grunewald, E.; Bergin, B.A.; Codd, S.L. Biofilm detection in natural unconsolidated porous media using a low-field magnetic resonance system. *Environ. Sci. Technol.* **2012**, *47*, 987–992. [CrossRef]
100. Lee, B.H.; Lee, S.K. Probing the water distribution in porous model sands with two immiscible fluids: A nuclear magnetic resonance micro-imaging study. *J. Hydrol.* **2017**, *553*, 637–650. [CrossRef]
101. Vogt, S.J.; Sanderlin, A.B.; Seymour, J.D.; Codd, S.L. Permeability of a growing biofilm in a porous media fluid flow analyzed by magnetic resonance displacement-relaxation correlations. *Biotechnol. Bioeng.* **2012**, *110*, 1366–1375. [CrossRef]
102. Carrel, M.; Beltran, M.A.; Morales, V.L.; Derlon, N.; Morgenroth, E.; Kaufmann, R.; Holzner, M. Biofilm imaging in porous media by laboratory X-ray tomography: Combining a non-destructive contrast agent with propagation-based phase-contrast imaging tools. *PLoS ONE* **2017**, *12*, e0180374. [CrossRef]
103. Downie, H.F.; Valentine, T.; Otten, W.; Spiers, A.; Dupuy, L. Transparent soil microcosms allow 3D spatial quantification of soil microbiological processes in vivo. *Plant Signal. Behav.* **2014**, *9*, e970421. [CrossRef]
104. Downie, H.; Holden, N.; Otten, W.; Spiers, A.; Valentine, T.; Dupuy, L.X. Transparent soil for imaging the rhizosphere. *PLoS ONE* **2012**, *7*, e44276. [CrossRef] [PubMed]
105. O'Callaghan, F.E.; Braga, R.; Neilson, R.; Macfarlane, S.A.; Dupuy, L.X. New live screening of plant-nematode interactions in the rhizosphere. *Sci. Rep.* **2018**, *8*, 1–17. [CrossRef]
106. Foster, R.C. Microenvironments of soil microorganisms. *Biol. Fertil. Soils* **1988**, *6*, 189–203. [CrossRef]
107. Lamandé, M.; Schjønning, P.; Ferro, N.D.; Morari, F. Soil pore system evaluated from gas measurements and CT images: A conceptual study using artificial, natural and 3D-printed soil cores. *Eur. J. Soil Sci.* **2021**, *72*, 769–781. [CrossRef]
108. Otten, W.; Pajor, R.; Schmidt, S.; Baveye, P.; Hague, R.; Falconer, R.E. Combining X-ray CT and 3D printing technology to produce microcosms with replicable, complex pore geometries. *Soil Biol. Biochem.* **2012**, *51*, 53–55. [CrossRef]
109. Aleklett, K.; Kiers, E.T.; Ohlsson, P.; Shimizu, T.S.; Caldas, V.E.; Hammer, E.C. Build your own soil: Exploring microfluidics to create microbial habitat structures. *ISME J.* **2018**, *12*, 312–319. [CrossRef]
110. Burmeister, A.; Grünberger, A. Microfluidic cultivation and analysis tools for interaction studies of microbial co-cultures. *Curr. Opin. Biotechnol.* **2020**, *62*, 106–115. [CrossRef]
111. Harvey, H.J.; Wildman, R.D.; Mooney, S.J.; Avery, S.V. Challenges and approaches in assessing the interplay between microorganisms and their physical micro-environments. *Comput. Struct. Biotechnol. J.* **2020**, *18*, 2860–2866. [CrossRef]
112. Wessel, A.K.; Hmelo, L.; Parsek, M.R.; Whiteley, M. Going local: Technologies for exploring bacterial microenvironments. *Nat. Rev. Genet.* **2013**, *11*, 337–348. [CrossRef]
113. Zhou, W.; Le, J.; Chen, Y.; Cai, Y.; Hong, Z.; Chai, Y. Recent advances in microfluidic devices for bacteria and fungus research. *TrAC Trends Anal. Chem.* **2019**, *112*, 175–195. [CrossRef]
114. Weibel, D.B.; DiLuzio, W.R.; Whitesides, G.M. Microfabrication meets microbiology. *Nat. Rev. Microbiol.* **2007**, *5*, 209–218. [CrossRef]
115. Seymour, J.R.; Ahmed, T.; Stocker, R. A microfluidic chemotaxis assay to study microbial behavior in diffusing nutrient patches. *Limnol. Oceanogr. Methods* **2008**, *6*, 477–488. [CrossRef]
116. Kim, H.J.; Boedicker, J.Q.; Choi, J.W.; Ismagilov, R.F. Defined spatial structure stabilizes a synthetic multispecies bacterial community. *Proc. Natl. Acad. Sci. USA* **2008**, *105*, 18188–18193. [CrossRef]
117. Boedicker, J.Q.; Vincent, M.E.; Ismagilov, R.F. Microfluidic confinement of single cells of bacteria in small volumes initiates high-density behavior of quorum sensing and growth and reveals its variability. *Angew. Chem. Int. Ed.* **2009**, *48*, 5908–5911. [CrossRef]
118. Mukhopadhyay, R. When PDMS isn't the best. What are its weaknesses, and which other polymers can researchers add to their toolboxes? *Anal. Chem.* **2007**, *79*, 3248–3253. [CrossRef] [PubMed]
119. Mcdonald, J.C.; Duffy, D.C.; Anderson, J.R.; Chiu, D.T. Review general fabrication of microfluidic systems in poly (dimethylsiloxane). *Electrophoresis* **2000**, *21*, 27–40. [CrossRef]
120. Connell, J.L.; Wessel, A.K.; Parsek, M.R.; Ellington, A.D.; Whiteley, M.; Shear, J.B. Probing prokaryotic social behaviors with bacterial "lobster traps". *mBio* **2010**, *1*, 1–8. [CrossRef] [PubMed]

121. Baca, H.K.; Ashley, C.; Carnes, E.; Lopez, D.; Flemming, J.; Dunphy, D.; Singh, S.; Chen, Z.; Liu, N.; Fan, H.; et al. Cell-directed assembly of lipid-silica nanostructures providing extended cell viability. *Science* **2006**, *313*, 337–341. [CrossRef] [PubMed]
122. Carnes, E.C.; Lopez, D.M.; Donegan, N.; Cheung, A.; Gresham, H.; Timmins, G.; Brinker, C.J. Confinement-induced quorum sensing of individual *Staphylococcus aureus* bacteria. *Nat. Chem. Biol.* **2009**, *6*, 41–45. [CrossRef] [PubMed]
123. Deng, J.; Orner, E.; Chau, J.F.; Anderson, E.M.; Kadilak, A.L.; Rubinstein, R.; Bouchillon, G.M.; Goodwin, R.; Gage, D.; Shor, L.M. Synergistic effects of soil microstructure and bacterial EPS on drying rate in emulated soil micromodels. *Soil Biol. Biochem.* **2015**, *83*, 116–124. [CrossRef]
124. Rubinstein, R.; Kadilak, A.L.; Cousens, V.C.; Gage, D.; Shor, L.M. Protist-facilitated particle transport using emulated soil micromodels. *Environ. Sci. Technol.* **2015**, *49*, 1384–1391. [CrossRef] [PubMed]
125. Wondraczek, L.; Pohnert, G.; Schacher, F.H.; Köhler, A.; Gottschaldt, M.; Schubert, U.S.; Küsel, K.; Brakhage, A.A. Artificial microbial arenas: Materials for observing and manipulating microbial consortia. *Adv. Mater.* **2019**, *31*, e1900284. [CrossRef]
126. Gimeno, A.; Stanley, C.E.; Ngamenie, Z.; Hsung, M.-H.; Walder, F.; Schmieder, S.S.; Bindschedler, S.; Junier, P.; Keller, B.; Vogelgsang, S. A versatile microfluidic platform measures hyphal interactions between *Fusarium graminearum* and *Clonostachys rosea* in real-time. *Commun. Biol.* **2021**, *4*, 1–10. [CrossRef] [PubMed]
127. Hanson, K.L.; Nicolau, D.V.; Filipponi, L.; Wang, L.; Lee, A.P. Fungi use efficient algorithms for the exploration of microfluidic networks. *Small* **2006**, *2*, 1212–1220. [CrossRef] [PubMed]
128. Tan, J.Y.; Wang, S.; Dick, G.J.; Young, V.B.; Sherman, D.H.; Burns, M.A.; Lin, X.N. Co-cultivation of microbial sub-communities in microfluidic droplets facilitates high-resolution genomic dissection of microbial 'dark matter'. *Integr. Biol.* **2020**, *12*, 263–274. [CrossRef]
129. Kaehr, B.; Shear, J.B. Multiphoton fabrication of chemically responsive protein hydrogels for microactuation. *Proc. Natl. Acad. Sci. USA* **2008**, *105*, 8850–8854. [CrossRef]
130. Zhang, Z.; Boccazzi, P.; Choi, H.-G.; Perozziello, G.; Sinskey, A.J.; Jensen, K. Microchemostat—Microbial continuous culture in a polymer-based, instrumented microbioreactor. *Lab Chip* **2006**, *6*, 906–913. [CrossRef]
131. Pamp, S.J.; Sternberg, C.; Tolker-Nielsen, T. Insight into the microbial multicellular lifestyle via flow-cell technology and confocal microscopy. *Cytom. Part A* **2008**, *75*, 90–103. [CrossRef]
132. Massalha, H.; Korenblum, E.; Malitsky, S.; Shapiro, O.H.; Aharoni, A. Live imaging of root–bacteria interactions in a microfluidics setup. *Proc. Natl. Acad. Sci. USA* **2017**, *114*, 4549–4554. [CrossRef]
133. Schmieder, S.S.; Stanley, C.; Rzepiela, A.; van Swaay, D.; Sabotič, J.; Nørrelykke, S.; Demello, A.J.; Aebi, M.; Künzler, M. Bidirectional propagation of signals and nutrients in fungal networks via specialized hyphae. *Curr. Biol.* **2019**, *29*, 217–228. [CrossRef] [PubMed]
134. Tayyrov, A.; Stanley, C.E.; Azevedo, S.; Künzler, M. Combining microfluidics and RNA-sequencing to assess the inducible defensome of a mushroom against nematodes. *BMC Genom.* **2019**, *20*, 1–13. [CrossRef]
135. Uehling, J.K.; Entler, M.R.; Meredith, H.R.; Millet, L.J.; Timm, C.M.; Aufrecht, J.A.; Bonito, G.M.; Engle, N.L.; Labbé, J.L.; Doktycz, M.J.; et al. Microfluidics and metabolomics reveal symbiotic bacterial–fungal interactions between *Mortierella elongata* and Burkholderia include metabolite exchange. *Front. Microbiol.* **2019**, *10*, 2163. [CrossRef] [PubMed]
136. Stanley, C.; Stöckli, M.; van Swaay, D.; Sabotič, J.; Kallio, P.T.; Künzler, M.; Demello, A.J.; Aebi, M. Probing bacterial–fungal interactions at the single cell level. *Integr. Biol.* **2014**, *6*, 935–945. [CrossRef]
137. Eun, Y.J.; Utada, A.S.; Copeland, M.F.; Takeuchi, S.; Weibel, D.B. Encapsulating bacteria in agarose microparticles using microfluidics for high-throughput cell analysis and isolation. *ACS Chem. Biol.* **2010**, *6*, 260–266. [CrossRef]
138. Churski, K.; Kaminski, T.S.; Jakiela, S.; Kamysz, W.; Barańska-Rybak, W.; Weibel, D.B.; Garstecki, P. Rapid screening of antibiotic toxicity in an automated microdroplet system. *Lab Chip* **2012**, *12*, 1629–1637. [CrossRef]
139. Abraham, W.-R. Applications and impacts of stable isotope probing for analysis of microbial interactions. *Appl. Microbiol. Biotechnol.* **2014**, *98*, 4817–4828. [CrossRef]
140. Partida-Martinez, L.P.; Groth, I.; Schmitt, I.; Richter, W.; Roth, M.; Hertweck, C. *Burkholderia rhizoxinica* sp. nov. and *Burkholderia endofungorum* sp. nov., bacterial endosymbionts of the plant-pathogenic fungus *Rhizopus microsporus*. *Int. J. Syst. Evol. Microbiol.* **2007**, *57*, 2583–2590. [CrossRef]
141. Söderström, B.; Erland, S. Isolation of fluorescein diacetate stained hyphae from soil by micromanipulation. *Trans. Br. Mycol. Soc.* **1986**, *86*, 465–468. [CrossRef]
142. Dennis, P.G.; Miller, A.J.; Clark, I.M.; Taylor, R.G.; Valsami-Jones, E.; Hirsch, P.R. A novel method for sampling bacteria on plant root and soil surfaces at the microhabitat scale. *J. Microbiol. Methods* **2008**, *75*, 12–18. [CrossRef] [PubMed]
143. Ashida, N.; Ishii, S.; Hayano, S.; Tago, K.; Tsuji, T.; Yoshimura, Y.; Otsuka, S.; Senoo, K. Isolation of functional single cells from environments using a micromanipulator: Application to study denitrifying bacteria. *Appl. Microbiol. Biotechnol.* **2009**, *85*, 1211–1217. [CrossRef] [PubMed]
144. Nishizawa, T.; Tago, K.; Uei, Y.; Ishii, S.; Isobe, K.; Otsuka, S.; Senoo, K. Advantages of functional single-cell isolation method over standard agar plate dilution method as a tool for studying denitrifying bacteria in rice paddy soil. *AMB Express* **2012**, *2*, 50. [CrossRef] [PubMed]
145. Fröhlich, J.; König, H. New techniques for isolation of single prokaryotic cells. *FEMS Microbiol. Rev.* **2000**, *24*, 567–572. [CrossRef]

146. Whitley, K.D.; Comstock, M.J.; Chemla, Y.R. High-resolution "fleezers": Dual-trap optical tweezers combined with single-molecule fluorescence detection. In *Optical Tweezers. Methods in Molecular Biology*; Gennerich, A., Ed.; Humana Press: New York, NY, USA, 2017; Volume 1486, pp. 183–256. [CrossRef]
147. Ringeisen, B.R.; Rincon, K.; Fitzgerald, L.A.; Fulmer, P.A.; Wu, P.K. Printing soil: A single-step, high-throughput method to isolate micro-organisms and near-neighbour microbial consortia from a complex environmental sample. *Methods Ecol. Evol.* **2014**, *6*, 209–217. [CrossRef]
148. Sato, Y.; Narisawa, K.; Tsuruta, K.; Umezu, M.; Nishizawa, T.; Tanaka, K.; Yamaguchi, K.; Komatsuzaki, M.; Ohta, H. Detection of betaproteobacteria inside the mycelium of the fungus *Mortierella elongata*. *Microbes Environ.* **2010**, *25*, 321–324. [CrossRef] [PubMed]
149. Ghodsalavi, B.; Svenningsen, N.B.; Hao, X.; Olsson, S.; Nicolaisen, M.H.; Abu Al-Soud, W.; Sørensen, S.J.; Nybroe, O. A novel baiting microcosm approach used to identify the bacterial community associated with *Penicillium bilaii* hyphae in soil. *PLoS ONE* **2017**, *12*, e0187116. [CrossRef] [PubMed]
150. Postma, J.; Altemüller, H.-J. Bacteria in thin soil sections stained with the fluorescent brightener calcofluor white M2R. *Soil Biol. Biochem.* **1990**, *22*, 89–96. [CrossRef]
151. Hoffman, M.T.; Gunatilaka, M.K.; Wijeratne, K.; Gunatilaka, L.; Arnold, A.E. Endohyphal bacterium enhances production of indole-3-acetic acid by a foliar fungal endophyte. *PLoS ONE* **2013**, *8*, e73132. [CrossRef] [PubMed]
152. Bertaux, J.; Schmid, M.; Prevost-Boure, N.C.; Churin, J.L.; Hartmann, A.; Garbaye, J.; Frey-Klett, P. In situ identification of intracellular bacteria related to *Paenibacillus* spp. in the mycelium of the ectomycorrhizal fungus *Laccaria bicolor* S238N. *Appl. Environ. Microbiol.* **2003**, *69*, 4243–4248. [CrossRef]
153. Hoffman, M.T.; Arnold, A.E. Diverse bacteria inhabit living hyphae of phylogenetically diverse fungal endophytes. *Appl. Environ. Microbiol.* **2010**, *76*, 4063–4075. [CrossRef]
154. Naumann, M.; Schüssler, A.; Bonfante, P. The obligate endobacteria of arbuscular mycorrhizal fungi are ancient heritable components related to the mollicutes. *ISME J.* **2010**, *4*, 862–871. [CrossRef] [PubMed]
155. Chen, A.Y.-Y.; Chen, A. Fluorescence in situ hybridization. *J. Investig. Dermatol.* **2013**, *133*, 1–4. [CrossRef] [PubMed]
156. Schmidt, H.; Eickhorst, T.; Mußmann, M. Gold-FISH: A new approach for the in situ detection of single microbial cells combining fluorescence and scanning electron microscopy. *Syst. Appl. Microbiol.* **2012**, *35*, 518–525. [CrossRef] [PubMed]
157. Schmidt, H.; Eickhorst, T.; Tippkötter, R. Evaluation of tyramide solutions for an improved detection and enumeration of single microbial cells in soil by CARD-FISH. *J. Microbiol. Methods* **2012**, *91*, 399–405. [CrossRef]
158. Schmidt, H.; Eickhorst, T. Detection and quantification of native microbial populations on soil-grown rice roots by catalyzed reporter deposition-fluorescencein situhybridization. *FEMS Microbiol. Ecol.* **2014**, *87*, 390–402. [CrossRef]
159. Behrens, S.; Lösekann, T.; Pett-Ridge, J.; Weber, P.K.; Ng, W.-O.; Stevenson, B.; Hutcheon, I.D.; Relman, D.A.; Spormann, A.M. Linking microbial phylogeny to metabolic activity at the single-cell level by using enhanced element labeling-catalyzed reporter deposition fluorescence in situ hybridization (EL-FISH) and NanoSIMS. *Appl. Environ. Microbiol.* **2008**, *74*, 3143–3150. [CrossRef] [PubMed]
160. Nunan, N.; Ritz, K.; Crabb, D.; Harris, K.; Wu, K.; Crawford, J.W.; Young, I.M. Quantification of the in situ distribution of soil bacteria by large-scale imaging of thin sections of undisturbed soil. *FEMS Microbiol. Ecol.* **2001**, *37*, 67–77. [CrossRef]
161. Hapca, S.; Baveye, P.; Wilson, C.; Lark, R.; Otten, W. Three-dimensional mapping of soil chemical characteristics at micrometric scale by combining 2D SEM-EDX data and 3D X-ray CT images. *PLoS ONE* **2015**, *10*, e0137205. [CrossRef] [PubMed]
162. Nunan, N.; Wu, K.; Young, I.M.; Crawford, J.W.; Ritz, K. Spatial distribution of bacterial communities and their relationships with the micro-architecture of soil. *FEMS Microbiol. Ecol.* **2003**, *44*, 203–215. [CrossRef]
163. Oburger, E.; Schmidt, H. New methods to unravel rhizosphere processes. *Trends Plant Sci.* **2016**, *21*, 243–255. [CrossRef] [PubMed]
164. Bradley, R.S.; Robinson, I.; Yusuf, M. 3D X-ray nanotomography of cells grown on electrospun scaffolds. *Macromol. Biosci.* **2016**, *17*, 1–8. [CrossRef]
165. Bouckaert, L.; van Loo, D.; Ameloot, N.; Buchan, D.; van Hoorebeke, L.; Sleutel, S. Compatibility of X-ray micro-computed tomography with soil biological experiments. *Soil Biol. Biochem.* **2013**, *56*, 10–12. [CrossRef]
166. Sturrock, C.; Woodhall, J.; Brown, M.; Walker, C.; Mooney, S.; Ray, R.V. Effects of damping-off caused by *Rhizoctonia solani* anastomosis group 2-1 on roots of wheat and oil seed rape quantified using X-ray computed tomography and real-time PCR. *Front. Plant Sci.* **2015**, *6*, 461. [CrossRef]
167. Hingley-Wilson, S.M.; Ma, N.; Hu, Y.; Casey, R.; Bramming, A.; Curry, R.J.; Tang, H.L.; Wu, H.; Butler, R.E.; Jacobs, W.R.; et al. Loss of phenotypic inheritance associated with ydcI mutation leads to increased frequency of small, slow persisters in *Escherichia coli*. *Proc. Natl. Acad. Sci. USA* **2020**, *117*, 4152–4157. [CrossRef] [PubMed]
168. Juyal, A.; Otten, W.; Baveye, P.C.; Eickhorst, T. Influence of soil structure on the spread of *Pseudomonas fluorescens* in soil at microscale. *Eur. J. Soil Sci.* **2021**, *72*, 141–153. [CrossRef]
169. Zappala, S.; Helliwell, J.R.; Tracy, S.; Mairhofer, S.; Sturrock, C.; Pridmore, T.; Bennett, M.; Mooney, S. Effects of X-ray dose on rhizosphere studies using X-ray computed tomography. *PLoS ONE* **2013**, *8*, e67250. [CrossRef] [PubMed]
170. Schmidt, H.; Vetterlein, D.; Köhne, J.M.; Eickhorst, T. Negligible effect of X-ray μ-CT scanning on archaea and bacteria in an agricultural soil. *Soil Biol. Biochem.* **2015**, *84*, 21–27. [CrossRef]
171. Arendt, K.R.; Hockett, K.; Araldi-Brondolo, S.J.; Baltrus, D.A.; Arnold, A.E. Isolation of Endohyphal bacteria from foliar ascomycota and in vitro establishment of their symbiotic associations. *Appl. Environ. Microbiol.* **2016**, *82*, 2943–2949. [CrossRef]

172. Benoit, I.; van den Esker, M.H.; Patyshakuliyeva, A.; Mattern, D.J.; Blei, F.; Zhou, M.; Dijksterhuis, J.; Brakhage, A.A.; Kuipers, O.P.; de Vries, R.P.; et al. *Bacillus subtilis* attachment to *Aspergillus nigerhyphae* results in mutually altered metabolism. *Environ. Microbiol.* **2014**, *17*, 2099–2113. [CrossRef]
173. Schuster, M.; Kilaru, S.; Guo, M.; Sommerauer, M.; Lin, C.; Steinberg, G. Red fluorescent proteins for imaging *Zymoseptoria tritici* during invasion of wheat. *Fungal Genet. Biol.* **2015**, *79*, 132–140. [CrossRef]
174. Lichius, A. Live-cell imaging in Trichoderma. In *New and Future Developments in Microbial Biotechnology and Bioengineering: Recent Developments in Trichoderma Research*; Gupta, V.K., Zeilinger, S., Singh, H.B., Druzhinina, I., Eds.; Elsevier: Amsterdam, The Netherlands, 2020; pp. 75–108.
175. Bard, A.J.; Mirkin, M.V. (Eds.) *Scanning Electrochemical Microscopy*; Marcel Dekker: New York, NY, USA, 2001.
176. Liu, X.; Ramsey, M.M.; Chen, X.; Koley, D.; Whiteley, M.; Bard, A.J. Real-time mapping of a hydrogen peroxide concentration profile across a polymicrobial bacterial biofilm using scanning electrochemical microscopy. *Proc. Natl. Acad. Sci. USA* **2011**, *108*, 2668–2673. [CrossRef]
177. Buchberger, A.R.; Delaney, K.; Johnson, J.; Jillian, J. Mass spectrometry imaging: A review of emerging advancements and future insights. *Anal. Chem.* **2018**, *90*, 240–265. [CrossRef]
178. Watrous, J.D.; Dorrestein, P.C. Imaging mass spectrometry in microbiology. *Nat. Rev. Genet.* **2011**, *9*, 683–694. [CrossRef] [PubMed]
179. Esquenazi, E.; Yang, Y.-L.; Watrous, J.; Gerwick, W.H.; Dorrestein, P.C. Imaging mass spectrometry of natural products. *Nat. Prod. Rep.* **2009**, *26*, 1521–1534. [CrossRef]
180. Yang, Y.-L.; Xu, Y.; Straight, P.; Dorrestein, P.C. Translating metabolic exchange with imaging mass spectrometry. *Nat. Chem. Biol.* **2009**, *5*, 885–887. [CrossRef] [PubMed]
181. Watrous, J.; Hendricks, N.; Meehan, M.; Dorrestein, P.C. capturing bacterial metabolic exchange using thin film desorption electrospray ionization-imaging mass spectrometry. *Anal. Chem.* **2010**, *82*, 1598–1600. [CrossRef] [PubMed]
182. Esquenazi, E.; Coates, C.; Simmons, L.; Gonzalez, D.; Gerwick, W.H.; Dorrestein, P.C. Visualizing the spatial distribution of secondary metabolites produced by marine cyanobacteria and sponges via MALDI-TOF imaging. *Mol. BioSyst.* **2008**, *4*, 562–570. [CrossRef] [PubMed]
183. Liu, W.-T.; Yang, Y.-L.; Xu, Y.; Lamsa, A.; Haste, N.M.; Yang, J.Y.; Ng, J.; Gonzalez, D.; Ellermeier, C.; Straight, P.; et al. Imaging mass spectrometry of intraspecies metabolic exchange revealed the cannibalistic factors of *Bacillus subtilis*. *Proc. Natl. Acad. Sci. USA* **2010**, *107*, 16286–16290. [CrossRef] [PubMed]
184. Holzlechner, M.; Reitschmidt, S.; Gruber, S.; Zeilinger, S.; Marchetti-Deschmann, M. Visualizing fungal metabolites during mycoparasitic interaction by MALDI mass spectrometry imaging. *Proteomics* **2016**, *16*, 1742–1746. [CrossRef]
185. Moree, W.J.; Yang, J.Y.; Zhao, X.; Liu, W.-T.; Aparicio, M.; Atencio, L.; Ballesteros, J.; Sanchez, J.; Gavilan, R.; Gutiérrez, M.; et al. Imaging mass spectrometry of a coral microbe interaction with fungi. *J. Chem. Ecol.* **2013**, *39*, 1045–1054. [CrossRef]
186. Boya, P.C.A.; Fernández-Marín, H.; Mejia, L.C.; Spadafora, C.; Dorrestein, P.C.; Gutiérrez, M. Imaging mass spectrometry and MS/MS molecular networking reveals chemical interactions among cuticular bacteria and pathogenic fungi associated with fungus-growing ants. *Sci. Rep.* **2017**, *7*, 5604. [CrossRef]
187. Lane, A.L.; Nyadong, L.; Galhena, A.S.; Shearer, T.L.; Stout, E.P.; Parry, R.M.; Kwasnik, M.; Wang, M.D.; Hay, M.E.; Fernandez, F.M.; et al. Desorption electrospray ionization mass spectrometry reveals surface-mediated antifungal chemical defense of a tropical seaweed. *Proc. Natl. Acad. Sci. USA* **2009**, *106*, 7314–7319. [CrossRef]
188. Pett-Ridge, J.; Weber, P.K. NanoSIP: NanoSIMS applications for microbial biology. In *Microbial Systems Biology. Methods in Molecular Biology (Methods and Protocols)*; Navid, A., Ed.; Humana Press: Totowa, NJ, USA, 2012; Volume 881, pp. 375–408. [CrossRef]
189. Vaidyanathan, S.; Fletcher, J.; Goodacre, R.; Lockyer, N.P.; Micklefield, J.; Vickerman, J.C. Subsurface biomolecular imaging of *Streptomyces coelicolor* using secondary ion mass spectrometry. *Anal. Chem.* **2008**, *80*, 1942–1951. [CrossRef]
190. Murrell, J.C.; Whiteley, A.S. *Stable Isotope Probing and Related Technologies*; American Society for Microbiology Press: Washington, DC, USA, 2010. [CrossRef]
191. Dumont, M.G.; Murrell, J.C. Stable isotope probing—Linking microbial identity to function. *Nat. Rev. Genet.* **2005**, *3*, 499–504. [CrossRef]
192. Rime, T.; Hartmann, M.; Frey, B. Potential sources of microbial colonizers in an initial soil ecosystem after retreat of an alpine glacier. *ISME J.* **2016**, *10*, 1625–1641. [CrossRef]
193. Pinto-Tomás, A.A.; Anderson, M.A.; Suen, G.; Stevenson, D.M.; Chu, F.S.T.; Cleland, W.W.; Weimer, P.J.; Currie, C.R. Symbiotic nitrogen fixation in the fungus gardens of leaf-cutter ants. *Science* **2009**, *326*, 1120–1123. [CrossRef]
194. Lee, N.; Nielsen, P.H.; Andreasen, K.H.; Juretschko, S.; Nielsen, J.L.; Schleifer, K.-H.; Wagner, M. Combination of fluorescent in situ hybridization and microautoradiography—A new tool for structure-function analyses in microbial ecology. *Appl. Environ. Microbiol.* **1999**, *65*, 1289–1297. [CrossRef]
195. Ouverney, C.C.; Fuhrman, J.A. Combined microautoradiography–16S rRNA probe technique for determination of radioisotope uptake by specific microbial cell types in situ. *Appl. Environ. Microbiol.* **1999**, *65*, 1746–1752. [CrossRef]
196. Adamczyk, J.; Hesselsoe, M.; Iversen, N.; Horn, M.; Lehner, A.; Nielsen, P.H.; Schloter, M.; Roslev, P.; Wagner, M. The isotope array, a new tool that employs substrate-mediated labeling of rRNA for determination of microbial community structure and function. *Appl. Environ. Microbiol.* **2003**, *69*, 6875–6887. [CrossRef] [PubMed]

197. Pilloni, G.; von Netzer, F.; Engel, M.; Lueders, T. Electron acceptor-dependent identification of key anaerobic toluene degraders at a tar-oil-contaminated aquifer by Pyro-SIP. *FEMS Microbiol. Ecol.* **2011**, *78*, 165–175. [CrossRef]
198. Eichorst, S.A.; Kuske, C.R. Identification of cellulose-responsive bacterial and fungal communities in geographically and edaphically different soils by using stable isotope probing. *Appl. Environ. Microbiol.* **2012**, *78*, 2316–2327. [CrossRef] [PubMed]
199. López-Mondéjar, R.; Brabcova, V.; Štursová, M.; Davidová, A.; Jansa, J.; Cajthaml, T.; Baldrian, P. Decomposer food web in a deciduous forest shows high share of generalist microorganisms and importance of microbial biomass recycling. *ISME J.* **2018**, *12*, 1768–1778. [CrossRef] [PubMed]
200. Sheik, A.R.; Brussaard, C.P.D.; Lavik, G.; Lam, P.; Musat, N.; Krupke, A.; Littmann, S.; Strous, M.; Kuypers, M.M.M. Responses of the coastal bacterial community to viral infection of the algae *Phaeocystis globosa*. *ISME J.* **2014**, *8*, 212–225. [CrossRef]
201. Musat, N.; Musat, F.; Weber, P.K.; Pett-Ridge, J. Tracking microbial interactions with NanoSIMS. *Curr. Opin. Biotechnol.* **2016**, *41*, 114–121. [CrossRef]
202. Dekas, A.E.; Poretsky, R.S.; Orphan, V.J. Deep-sea archaea fix and share nitrogen in methane-consuming microbial consortia. *Science* **2009**, *326*, 422–426. [CrossRef]
203. Orphan, V.J.; House, C.H.; Hinrichs, K.-U.; McKeegan, K.D.; DeLong, E.F. Methane-consuming archaea revealed by directly coupled isotopic and phylogenetic analysis. *Science* **2001**, *293*, 484–487. [CrossRef]
204. Tai, V.; Carpenter, K.J.; Weber, P.K.; Nalepa, C.A.; Perlman, S.J.; Keeling, P.J. Genome evolution and nitrogen fixation in bacterial ectosymbionts of a protist inhabiting wood-feeding cockroaches. *Appl. Environ. Microbiol.* **2016**, *82*, 4682–4695. [CrossRef] [PubMed]
205. Woebken, D.; Burow, L.C.; Prufert-Bebout, L.; Bebout, B.M.; Hoehler, T.M.; Pett-Ridge, J.; Spormann, A.M.; Weber, P.K.; Singer, S.W. Identification of a novel cyanobacterial group as active diazotrophs in a coastal microbial mat using NanoSIMS analysis. *ISME J.* **2012**, *6*, 1427–1439. [CrossRef] [PubMed]
206. Fike, D.A.; Gammon, C.L.; Ziebis, W.; Orphan, V.J. Micron-scale mapping of sulfur cycling across the oxycline of a cyanobacterial mat: A paired nanoSIMS and CARD-FISH approach. *ISME J.* **2008**, *2*, 749–759. [CrossRef] [PubMed]
207. Alonso, C.; Musat, N.; Adam, B.; Kuypers, M.; Amann, R. HISH–SIMS analysis of bacterial uptake of algal-derived carbon in the Río de la Plata estuary. *Syst. Appl. Microbiol.* **2012**, *35*, 541–548. [CrossRef] [PubMed]
208. Majchrzak, T.; Wojnowski, W.; Lubinska-Szczygeł, M.; Różańska, A.K.; Namieśnik, J.; Dymerski, T. PTR-MS and GC-MS as complementary techniques for analysis of volatiles: A tutorial review. *Anal. Chim. Acta* **2018**, *1035*, 1–13. [CrossRef] [PubMed]

*Review*

# Thermodynamics of Soil Microbial Metabolism: Applications and Functions

**Nieves Barros**

Department of Applied Physics, University of Santiago de Compostela, 15782 Santiago de Compostela, Spain; nieves.barros@usc.es

**Featured Application: Energy rules life. All living systems keep themselves alive by balancing the energy input and output by universal thermodynamic principles. Soils are not an exception to this; however, their extraordinary complexity makes them poorly described as a thermodynamic system. This review shows how thermodynamics can be applied to study the role of the microbial community to keep the soil alive.**

**Abstract:** The thermodynamic characterization of soils would help to study and to understand their strategies for survival, as well as defining their evolutionary state. It is still a challenging goal due to difficulties in calculating the thermodynamic state variables (enthalpy, Gibbs energy, and entropy) of the reactions taking place in, and by, soils. Advances in instrumentation and methodologies are bringing options for those calculations, boosting the interest in this subject. The thermodynamic state variables involve considering the soil microbial functions as key channels controlling the interchange of matter and energy between soil and the environment, through the concept of microbial energy use efficiency. The role of microbial diversity using the energy from the soil organic substrates, and, therefore, the who, where, with whom, and why of managing that energy is still unexplored. It could be achieved by unraveling the nature of the soil organic substrates and by monitoring the energy released by the soil microbial metabolism when decomposing and assimilating those substrates. This review shows the state of the art of these concepts and the future impact of thermodynamics on soil science and on soil ecology.

**Keywords:** thermodynamics; soil; microbial metabolism; microbial diversity

**Citation:** Barros, N. Thermodynamics of Soil Microbial Metabolism: Applications and Functions. *Appl. Sci.* **2021**, *11*, 4962. https://doi.org/10.3390/app11114962

Academic Editors: Maraike Probst and Judith Ascher-Jenull

Received: 27 April 2021
Accepted: 26 May 2021
Published: 28 May 2021



## 1. Introduction

Soil is one of the main primary resources on earth, together with water. Both act together, playing an essential role in our survival. Soil is the main source of nutrients for living systems, and acts as a platform supporting structures for those living systems, at macro- and micro-scale, from humans to microorganisms. Soil science has been closely attached to human activity for those reasons. From the introduction of agriculture in the Neolithic to our days, the requirement for knowledge about soil has been constant.

Nowadays, soil research continues to be a vast multidisciplinary research area. In our era, our coexistence with a global climate change process has shown the scarcity of knowledge about the impact of temperature on soil fertility and soil structure. Temperature is one of the variables affecting thermodynamic state functions. When the interest in temperature on soil arises, the immediate addition of the term "thermodynamics" is unavoidable to discover next that soil is still not characterized as such. The fact that thermodynamics of soil systems is a vastly unexplored area is boosting the interest in the subject, because of important applications, such as controlling the role of temperature on soil fertility, characterizing the maturity state of soil ecosystems, or predicting their evolution [1,2]. The role of thermodynamics in ecology is not new and has been a matter of concern and development since the beginning of the last century, yielding not only relevant and high-impact, but also controversial, publications [3,4].

What is new now is that technology and the development of different methodologies make it possible to go beyond the existing thermodynamic theoretical approaches to soil ecology by calculating the missing thermodynamic state variables for soil reactions until recently: enthalpy, Gibbs energy, and entropy changes. This opens the possibility to apply them for all the biotic and abiotic reactions taking place in soils to achieve thermodynamic soil characterization. The task is large. Therefore, it can be useful for future work to center the current situation about soil thermodynamics and to focus on what we need, what we have, and how to do it.

Soil is an extraordinary complex media, thermodynamically considered as an open system interchanging matter and energy with the environment. The interchange of the matter is responsible for nutrient cycling and has been investigated since the 19th century [5]. Most of the soil chemical and biochemical studies focus on the elemental composition, such as carbon content, C; nitrogen, N, and other mass products from its decomposition, such as $CO_2$. C and $CO_2$ have been widely used for settling soil as a carbon sink and as a source of $CO_2$ to the atmosphere by different mass balances [6]. The interchange of matter takes place by different abiotic and biotic reactions. The biotic part involves a high number of different microorganisms, such as bacteria, fungi, and yeast, working as drivers in the interchange of different nutrients through microbial metabolic reactions. Without microorganisms, soil becomes an inorganic substrate incapable of sustaining life on Earth.

Soil microbiomes need different substrates feeding their metabolisms. On a mass basis, those soil substrates act as suppliers of single chemical elements through complex molecules constituting the soil organic matter, SOM. The chemical characterization of SOM is a challenge facing both technological and methodological limitations linked to the high chemical and physical complexity of SOM, and the lack of knowledge about SOM chemical and biochemical transformations. This is because most of the existing knowledge is published and spread across different areas of knowledge with poor contact among them [7–9].

Along the last century, soil began to be considered as a source of energy, too, and the soil microbial metabolism (SMM) began to be monitored on heat basis [10,11]. The development of highly sensitive isothermal calorimeters makes it possible to quantify the heat released by SMM [12] and to consider soil as a thermodynamic system where SOM is the reservoir of energy fueling the soil microbial community. Therefore, we have the two main ingredients to develop soil thermodynamics: mass and energy. We also have all the bioenergetics developed for microbial metabolism along the end of the 19th century and throughout the entire 20th century until now [13–15].

What we need for the thermodynamic characterization of soil is to connect the SOM energy budget to the dissipation of that energy by SMM through the thermodynamic state variables. It is essential to be able to calculate them for the SOM and for the SMM responsible for SOM decomposition. It is challenging, but not impossible now.

## 2. Materials and Methods

### 2.1. Thermodynamic Characterization of SOM

This involves determining the enthalpies of formation and combustion of SOM, $\Delta_f H_{SOM}$ and $\Delta_c H_{SOM}$, the Gibbs energy change of formation and combustion of SOM as well, $\Delta_f G_{SOM}$ and $\Delta_c G_{SOM}$, and their respective entropy changes, $\Delta S$. Their calculation involves writing the reactions for SOM formation and/or SOM combustion. There are two different options for this: the stoichiometric methods and the enthalpy models linking the energy of a substrate with its degree of reduction and/or degree of oxidation.

The stoichiometric models implicate the chemical formulation of substrates and reactants. There are some attempts towards SOM chemical formulation, but it is a complex and still challenging objective [16,17], especially when it involves the characterization of all the reactions that take place in the soil. The latter is interesting and necessary from a chemical perspective but may not give answers and understanding to important questions for soil and ecology research, such as the measurement of SOM recalcitrance [18], assessment about the evolution of soil ecosystems based on how energy and exergy is managed [19], the state of soil fertility, and the connection with soil microbial diversity [1]. These goals involve studying the thermodynamic variables from a more global perspective than analyzing individual soil chemical reactions. Additional methodological alternatives providing more data and global perspectives for interpreting the thermodynamics of soil biogeochemical reactions should be welcomed [20].

Some additional options could include the direct quantification of the energy budget of SOM, and the heat released by the microbial decomposition of that SOM. There is technology to measure the energy of organic substrates and thermodynamic models connecting this energy to their chemical composition by the degree of reduction and/or oxidation. That is, both SOM chemical formulation [16] and SOM energy content lead to the redox state of SOM [20,21]. The energy can be measured by bomb calorimetry and thermal analysis. Both methods yield the heat of combustion of organic substrates. Bomb calorimetry is considered as the standard methodology to obtain the heat capacity and enthalpy of combustion of organic substrates [22], but it is not efficient when applied to mineral soil samples for different reasons, such as incomplete combustion [22,23].

The other option is the simultaneous application of thermogravimetry (TG) and differential scanning calorimetry (DSC). Both methods have been applied in different soil studies, but not for yielding enthalpies of combustion until recently. Reasons are complex and linked to the design and evolution of thermal analysis. Until recently, most DSC devices were not designed for working with samples with high energy content. DSC experiments missed most of the energy from the organic substrates. The evolution of these devices towards simultaneous TG and DSC experiments with soils and organic samples are giving heat of combustion data close to those accepted for organic substrates [24]. Nevertheless, it is still necessary to settle a procedure for accurate measurements of the heat of combustion of soils by simultaneous TG–DSC. It would be desirable to make comparisons of the heat of combustion values with proximate analysis, as reported recently [25]. These procedures are in the process of development, turning approaches to the heat of combustion of SOM into a realistic goal.

By now, TG–DSC has allowed linking of the energy from SOM combusted by airflow in the DSC to the soil mass lost during the combustion. These experimental phases are properly explained by the literature giving that heat in kJ $g^{-1}$ OM. Software in the TG–DSC is not well-designed yet for this purpose, and to measure the heat of combustion of SOM is necessary to export TG–DSC data to external auxiliary software to perform integrations, derivatives, and adjustments of baselines manually.

For comparisons and applications of thermodynamic models, it is necessary to normalize the energy obtained in kJ $g^{-1}$ OM to the C content of the soil in C mole. That involves performing elemental analysis of the soil samples to relate the OM content obtained by the TG to the total C and/or organic C given by the elemental analysis [26]. To settle the correlation between C and SOM is essential for accurate normalizations since the existence of clay and/or carbonates may overlap with the OM content given by the TG measurements. That single correlation may optimize the most adequate value relating C to SOM in our samples to be used as the unit conversion factor. The high variety of soil properties could make this previous step essential for every soil, and it may be difficult to provide a general conversion factor. An example with some soils is shown in the results section of this paper. The soil samples were collected from the soil surface and at 5 cm of depth, representing Eutric vertisols and Podzols under oak mature forests collected in Ireland [26] and in the UK.

Calculation of the enthalpy of combustion of SOM is possible for the reaction taking place in the DSC (and also in a bomb calorimeter), summarized as follows:

$$SOM(s) + xO_2(g) = yCO_2(g) + H_2O(g) - \Delta_c H_{SOM} \tag{1}$$

The TG–DSC analysis must be done under a flow of dry air, as reported [25]. $\Delta_c H_{SOM}$ is directly related to the degree of reduction of organic substrates by well-known relations [27,28] involving the oxycaloric quotient, $Q_0$, by the following general equation:

$$\Delta_c H^0 = Q_0 \, \gamma_C \quad (2)$$

$\Delta_c H^0$ is the enthalpy of combustion of any organic substrate at standard conditions; $Q_0$ is the oxycaloric quotient representing the ratio between the enthalpy of combustion and the degree of reduction of C, $\gamma_C$, from organic substrates.

$Q_0$ values vary in literature from $-104$ to $-118$ kJ mol$^{-1}$ degree of reduction$^{-1}$; all of them assigned to various authors [21]. They were determined for different organic substrates first and for microbial biomass later, yielding slopes in a similar range [21]. The reason for those differences comprises the structure of molecules which were considered in the latest corrections reported [28]. Recent works implementing these concepts [26] used the Sandler and Orbey value of $-109$ kJ/C mol$^{-1}$ degree of reduction$^{-1}$ [28], but the reality is that the different values exhibited by the literature have not been applied for soils to compare results obtained by the different $Q_0$ values reported.

Roels, and Sandler and Orbey's correlations [27,28] give the Gibbs energy change for Equation (1), $\Delta_c G_{SOM}$. It would be interesting to apply both models to analyze and compare the resulting values for soil samples.

The entropy change of Equation (1) is an interesting goal too, because of its involvement in the evolution of soil ecosystems [29]. It can be determined by models such as the one proposed by Battley in 1999 [30]. It focuses on the application of the Hess law to microbial growth reactions where reactants and products are well-known. It is difficult to apply it when considering SOM as a reactant because we do not know the products for SOM biodecomposition in many cases. They can be assumed and summarized through different general concepts [31], but the reality is that we only can approach the formulation as done by the correlations between energy and the degree of reduction.

Another alternative for approaching entropy changes is the equation for the Gibbs energy for irreversible reactions, adapted to Equation (1):

$$\Delta_c G_{SOM} = \Delta_c H_{SOM} - T\Delta_c S_{SOM} \quad (3)$$

In this paper it is shown, as an example, the complete thermodynamic characterization of SOM for the set of soil samples collected from different depths mentioned before.

### 2.2. *Thermodynamics of the Soil Microbial Metabolism (SMM)*

Microorganisms decompose SOM by different biochemical paths and different electron acceptors [31]. Nevertheless, most of the studies focusing on SMM bioenergetics apply to the aerobic decomposition of SOM, where $O_2$ is the electron acceptor. The interest in characterizing SMM from a thermodynamic perspective is linked to the development of calorimeters that measure the heat released by SMM [32]. As the aerobic decomposition of SOM releases $CO_2$ too, calorimeters have been adapted to monitor the heat and $CO_2$ from SMM by different calorespirometric procedures [33]. These studies for soils started at the beginning of this century, involving very recent findings [34,35]. The concomitant measurements of heat and $CO_2$ have the advantage to yield the calorespirometric ratio (CR) of SMM, a metabolic indicator providing additional information about the nature of substrates from SOM being decomposed during the calorespirometric measurement [31,36]. CR is a measure of the enthalpy of those substrates [26] and the microbial metabolic C-use efficiency, CUE [37].

For the thermodynamic characterization of SMM, it is essential to link CR values to the enthalpy of the substrates [26,31]. Assuming CR represents the enthalpy for substrates being metabolically decomposed, $\Delta_r H_{SOM}$, it would be possible to yield the Gibbs energy change for microbial decomposition of SOM, $\Delta_r G_{SOM}$, by the models developed for microbial metabolism [20,27,28]. Comparing $\Delta_r G_{SOM}$ values with those from SOM combustion in Equation (1), $\Delta_c G_{SOM}$, can shed light on microbial strategies to keep soil ecosystems far from equilibrium, an unexplored field. It is also not known how the $\Delta_c H_{SOM}$ and $\Delta_c G_{SOM}$ values may influence the soil microbial functions and soil microbial diversity. It would be possible to localize soils with SOM at more, or less, degree of reduction by their $\Delta_c H_{SOM}$ and $\Delta_c G_{SOM}$ values, to relate them to the complexity of the SOM macromolecule and to the properties of the soil microbial community on those locations in terms of soil microbial diversity and soil microbial metabolic diversity. There is no previous work to this respect and there is an example of it in the results section.

The set of soil samples from different depths collected in oak mature forests were calorespirometrically characterized for aerobic decomposition of SOM by a microbial community at steady-state metabolism.

The reaction for SOM biodecomposition in this case is summarized as follows:

$$SOM(s) + O_2(g) = Products(s) + H_2O(l) + CO_2(g) - \Delta_r H_{SOM} \quad (4)$$

where $\Delta_r H_{SOM}$ is directly determined by calorespirometry as the CR. The experimental procedure is well-explained by the literature [38]. $\Delta_r H_{SOM}$ gives the degree of reduction, $\gamma_r$, of the substrates being metabolically decomposed [26]:

$$CR \sim \Delta_r H_{SOM} = \gamma_r / 4\,(-455)\ kJmol^{-1}O_2 \quad (5)$$

The Gibbs energy change for Equation (4), $\Delta_r G_{SOM}$, is given by the relations reported by Roels, and Sandler and Orbey.

The entropy change is determined by adaptation of Equation (3) to Equation (4) at the temperature of the calorespirometric measurement (298 K).

## 3. Results

### 3.1. Thermodynamic Characterization of SOM

Table 1 shows the elemental composition, C, H, N, of the soil samples collected for this study. LF represents samples from the soil surface, while M corresponds to the mineral soils collected at 5 cm depth from the soil surface. SOM is the percentage determined by TG. The C content of samples, as well as SOM, depletes from the soil surface to the mineral samples. Moreover, there are some differences among sampling sites, despite all of them representing mature oaks ecosystems at similar environmental conditions.

**Table 1.** Elemental composition of soil samples selected for this review. SOM percentages are determined by thermogravimetry (TG). LF indicates samples from the soil surface representing organic matter at a low degree of decomposition. M represents mineral soil samples taken in the same places as LF but at 5 cm depth from surface, where SOM is more transformed than LF layers. Samples ROG, BW, and NF come from the Alice Holt Research Station in the southeast of the UK. Samples DC, G, and K are from the southwest of Ireland and were used in previous work [26]. All of them represent mature forest oak ecosystems.

| Samples | $C_{tot}$ (%) | $C_{org}$ (%) | H (%) | N (%) | SOM (%) |
|---|---|---|---|---|---|
| ROG LF | 39 | 34 | 5.0 | 1.6 | 73 |
| BW LF | 46 | 41 | 6.5 | 2.5 | 81 |
| NF LF | 44 | 42 | 6.5 | 2.0 | 80 |
| DC LF | 50 | 42 | 5.5 | 1.4 | 95 |
| G LF | 51 | 46 | 5.3 | 1.7 | 97 |
| K LF | 51 | 40 | 6.4 | 1.7 | 96 |
| ROG M | 5 | 5 | 0.3 | 0.3 | 10 |
| BWM | 11 | 9 | 1.0 | 0.6 | 23 |
| NFM | 12 | 8 | 1.4 | 0.5 | 16 |
| DCM | 6 | 4 | 0.7 | 0.3 | 7 |
| G M | 5 | 4 | 0.5 | 0.2 | 9 |
| KM | 10 | 10 | 1.4 | 0.5 | 21 |

Correlations among these components can yield information about the SOM composition, indicating the percentages of N and H attached to C and SOM to weigh the contribution of inorganic material. In this particular case, the interest relies on the correlation between C and SOM to obtain the conversion factor from grams of SOM to carbon mole to be compared to the individual C/SOM relation for each sample.

Correlation among $C_{tot}$, $C_{org}$, and SOM gives r values of 0.98 at $p < 0.001$ in all cases and equations $C_{tot} = 0.530SOM + 0.646$ and $C_{org} = 0.462SOM + 0.172$, respectively. It will be used the conversion factor of 0.462 g C $g^{-1}$ SOM.

The soil thermal properties are shown in Table 2. The heat of combustion, $Q_{SOM}$, is obtained directly by DSC in kJ$g^{-1}$ OM, and corrected to yield the enthalpy of combustion of SOM, $\Delta_c H_{SOM}$, in kJ $mol^{-1}$ C, as shown in Figure 1.

**Table 2.** Thermodynamic data of samples obtained directly from DSC curves, $Q_{SOM}$ and $\Delta_c H_{SOM}$, and the degree of reduction of samples, $\gamma_{SOM}$ determined by Roels, $\gamma_{SOM}$R, and Sandler and Orbey, $\gamma_{SOM}$S&O, correlations for the LF soil layer and mineral soil samples, M. Samples ROG, BW, and NF are from the southeast of the UK. Samples DC, G, and K are from the southwest of Ireland. The reproducibility of measurements by DSC–TG is 5%. The uncertainty averaged for the degree of reduction is ±0.2 based on the standard errors reported for both models [27,28].

| Samples | $-Q_{SOM}$ kJ $g^{-1}$SOM LF | $-Q_{SOM}$ kJ $g^{-1}$SOM M | $-\Delta_c H_{SOM}$ kJ $mol^{-1}$C LF | $-\Delta_c H_{SOM}$ kJ $mol^{-1}$C M |
|---|---|---|---|---|
| ROG | 15.4 | 21.9 | 449 | 617 |
| BW | 15.1 | 20.7 | 441 | 586 |
| NF | 15.1 | 22.2 | 440 | 624 |
| DC | 15.0 | 26.4 | 437 | 734 |
| G | 16.4 | 25.0 | 475 | 697 |
| K | 15.9 | 21.3 | 461 | 602 |
| **Samples** | **$\gamma_{SOM}$R LF** | **$\gamma_{SOM}$S&O LF** | **$\gamma_{SOM}$R M** | **$\gamma_{SOM}$S&O M** |
| ROG | 3.90 | 4.11 | 5.36 | 5.66 |
| BW | 3.83 | 4.05 | 5.10 | 5.39 |
| NF | 3.83 | 4.04 | 5.43 | 5.72 |
| DC | 3.80 | 4.01 | 6.38 | 6.73 |
| G | 4.13 | 4.36 | 6.06 | 6.39 |
| K | 4.01 | 4.23 | 5.23 | 5.22 |

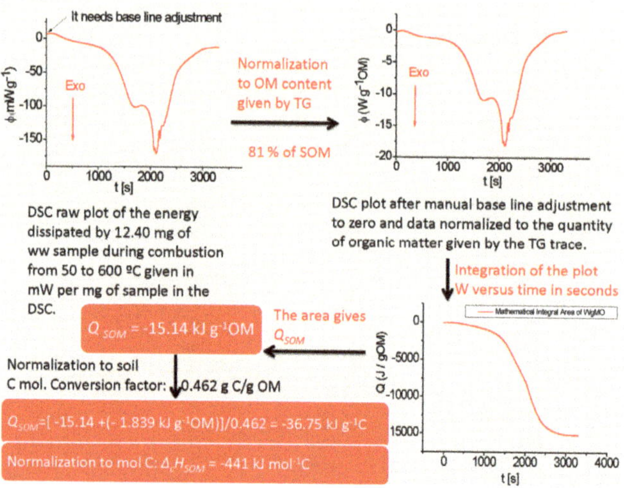


**Figure 1.** A summary of the procedure followed to approach the enthalpy of combustion of SOM, $\Delta_c H_{SOM}$, by simultaneous DSC–TG.

The Gibbs energy change is determined by the Roels and Sander and Orbey correlations, and the entropy change is from Equation (3).

$Q_{SOM}$ and $\Delta_c H_{SOM}$ values in Table 2 show higher values in mineral samples than in LF samples, indicating SOM at a higher degree of reduction as soil depth increases. This involves a change in SOM nature, and it is expected that it affects the soil microbial population, too.

The degree of reduction values yields the Gibbs energy change of SOM combustion, $\Delta_c G_{SOM}$, and, indirectly, the entropy change, $\Delta_c S_{SOM}$, for Equation (1). Results are shown in Table 3. The Gibbs energy becomes more negative in mineral samples than LF samples, independently of the models used. Although, apparently, these models yield similar values for Gibbs energy, incongruences between them appear when calculating the entropy changes. By comparing results using the Sandler and Orbey model alone [28], it is observed that a higher degree of reduction of SOM could yield a relative increment in the entropy change, compatible with the higher structural complexity of SOM and/or higher stable material in M samples than in LF samples. Little variability of the entropy change is also obtained among the different locations.

**Table 3.** Values of the Gibbs energy change of reaction (Equation (1)) obtained by applying the Roels correlation, $-\Delta_c G_{SOM}R$, and Sandler and Orbey correlation, $-\Delta_c G_{SOM}S\&O$, for LF and mineral, M, samples, together with their respective entropy changes for Reaction (1). The residual standard error for calculation of $-\Delta_c G_{SOM}R$ data is that reported by Roels, 18 kJ mol$^{-1}$, and that for estimation of $-\Delta_c G_{SOM}S\&O$ is 21.5 kJmol$^{-1}$ [26,27]. Samples ROG, BW, and NF are from the southeast of the UK. Samples DC, G, and K are from the southwest of Ireland.

| Samples | $-\Delta_c G_{SOM}R$ kJ mol$^{-1}$C LF | $-\Delta_c G_{SOM}S\&O$ kJ mol$^{-1}$C LF | $-\Delta_c G_{SOM}R$ kJ mol$^{-1}$C M | $-\Delta_c G_{SOM}S\&O$ kJ mol$^{-1}$C M |
|---|---|---|---|---|
| ROG | 455 | 454 | 594 | 624 |
| BW | 448 | 446 | 568 | 593 |
| NF | 448 | 445 | 599 | 631 |
| DC | 445 | 442 | 689 | 742 |
| G | 477 | 481 | 659 | 705 |
| K | 465 | 466 | 581 | 609 |
| **Samples** | **$\Delta_c S_{SOM}R$ J K$^{-1}$mol$^{-1}$C LF** | **$\Delta_c S_{SOM}S\&O$ J K$^{-1}$mol$^{-1}$C LF** | **$\Delta_c S_{SOM}R$ J K$^{-1}$mol$^{-1}$C M** | **$\Delta_c S_{SOM}S\&O$ J K$^{-1}$mol$^{-1}$C M** |
| ROG | 20.1 | 16.8 | −79.3 | 23.5 |
| BW | 23.5 | 16.8 | −60.4 | 23.5 |
| NF | 26.8 | 16.8 | −83.9 | 23.5 |
| DC | 26.9 | 16.8 | −151.0 | 26.9 |
| G | 6.7 | 20.1 | −127.5 | 26.9 |
| K | 13.4 | 16.8 | −70.5 | 23.5 |

### *3.2. Thermodynamics of SOM Microbial Decomposition*

Different calorimetric procedures address the bioenergetics of SOM decomposition. Some of the main goals are the thermodynamic characterization of the biochemical reactions involved in SMM [26,31], quantification of microbial metabolic efficiency [38–40], monitoring the biodegradation of external organic sources by different microbial metabolic paths [41], and its connection with microbial diversity [42].

Microbial diversity involves characterizing the composition of the microbial community and studies about microbial metabolic diversity. Although both are interesting for better characterization of soil biological properties, focusing on metabolic diversity could be more useful if the goal is SOM biodegradation. This is because biodecomposition is mainly ruled by enzyme diversity that has not necessarily been linked to higher microbial composition diversity. Diverse microorganisms contain the same enzymatic machinery. For this reason, higher diversity in microbial populations may not be reflected in biodecomposition rates. In most cases, these studies are still under development and there is little information about the role of microbial diversity on thermodynamics, metabolic efficiency, and biodegradability. This review addresses some of the existing results covering these topics.

3.2.1. Thermodynamic State Variables and Soil Microbial Diversity

The study of SOM biodecomposition by thermodynamic state variables determined experimentally remains a distinct goal [26,43,44]. This subject has been discussed typically on a theoretical basis [2,45] due to difficulties in calculating the thermodynamic state variables for SMM.

Differences in soil properties and soil chemical and thermal composition yield distinct microbial structures affecting the metabolic heat rate and the kinetics of decomposition of external C sources. This is reflected in the profiles of the calorimetric plots [41,43] that can be related to the decomposition of distinct substrates, but not to the genetic microbial structure in soils. It is demonstrated by calorimetry that different microbial structures may yield different metabolic efficiencies and that the CR may be sensitive to changes in the soil microbial composition linked to a certain soil management [39]. Some results also indicate how microbial diversity is involved in the sensitivity of the soils to temperature by calorimetry [46,47], but there are difficulties in relating all of them to a certain microbial structure because heat rates largely depend more on the enzymatic microbial diversity than on the microbial composition. That is, diverse microorganisms contributing to microbial genetic diversity can synthesize the same enzymes and run the same metabolic paths by similar metabolic heat rates. In this sense, calorimetry would be more useful for studying soil microbial metabolic diversity because it detects soil microbial biodegradation and assimilation of different substrates constituting a method for quantitative assessment of soil microbial metabolic diversity [41,43].

Nevertheless, none of those applications connects soil microbial diversity with thermodynamic functions.

This subsection provides some initial results connecting the thermodynamic characterization of SMM with the SOM thermodynamic properties shown in Tables 2 and 3. There are some additional samples to reveal the possible connection between the thermodynamic properties and the soil microbial metabolic diversity in mineral soil samples.

The enthalpy change of the microbial reaction taking place in a calorimeter can be directly determined by calorespirometry and the CR which gives the thermodynamic state variables for Equation (4). Table 4 shows the results obtained for the soil samples handled in this review. In this case, only Sandler and Orbey's model is applied to obtain Gibbs energy. The reason is to avoid incongruences with the entropy change, as happened with the characterization of SOM.

**Table 4.** Values for the calorespirometric ratio, CR, for the LF and M samples from the UK (ROG, BW, and NF) and Ireland (DC, G, and K); the degree of reduction of substrates being metabolized, $\gamma_r$, is obtained By Equation (5). Assuming CR represents the enthalpy change of Reaction (4), $\Delta_r H_{SOM}$; Gibbs energy change for Reaction (4) is also obtained by Sandler and Orbey's correlation, $\Delta_r G_{SOM}$, as well as entropy change for Reaction (4), obtained by Equation (3), $\Delta_r S_{SOM}$. Uncertainties for the degree of reduction and Gibbs energy change estimated from Sandler and Orbey's correlation are the same as reported in Tables 2 and 3. Standard deviation in CR samples is determined from two replicates of each sample.

| Samples | −CR kJ mol$^{-1}$$CO_2$-C LF | −CR kJ mol$^{-1}$$CO_2$-C M | $\gamma$ LF | $\gamma$ M | $-\Delta_r G_{SOM}$ kJmol$^{-1}$C LF | $-\Delta_r G_{SOM}$ kJmol$^{-1}$C M | $\Delta_r S_{SOM}$ JK$^{-1}$mol$^{-1}$C LF | $\Delta_r S_{SOM}$ JK$^{-1}$mol$^{-1}$C M |
|---|---|---|---|---|---|---|---|---|
| ROG | 423 ± 30 | 280 ± 7 | 3.88 | 2.57 | 427 | 283 | 16.0 | 10.6 |
| BW | 507 ± 19 | 276 ± 9 | 4.65 | 2.53 | 513 | 279 | 19.2 | 10.5 |
| NF | 312 ± 10 | 253 ± 12 | 2.86 | 2.32 | 316 | 256 | 11.8 | 9.6 |
| DC | 310 ± 44 | 400 ± 68 | 2.85 | 3.67 | 315 | 405 | 11.8 | 15.2 |
| G | 327 ± 41 | 485 ± 9 | 3.00 | 4.45 | 332 | 490 | 12.4 | 18.4 |
| K | 355 ± 51 | 421 ± 17 | 3.26 | 3.87 | 359 | 427 | 13.4 | 16.0 |

Results in Table 4 show changes in the CR from LF to M samples and among samples from different sites. Those differences can be explained by, or assumed as, changes in microbial metabolism. LF samples from the same site in the UK (ROG, BW, and NF) show decomposition of substrates at higher, the same, and lower degree of reduction than carbohydrates. All LF samples from separate sites in Ireland have similar CR values in the range reported for carbohydrate catabolism. Mineral samples from the UK show lower and more stable CR values than LF samples, suggesting a change in the microbial metabolism, characterized now by biodecomposition of substrates more oxidized than carbohydrates or by a higher component of anaerobic metabolism as depth increases. Mineral samples from Ireland show higher CR values than the LF ones, but in the carbohydrate range in all cases.

Gibbs energy follows the same trend as that of CR since the entropy change contributes little to Gibbs energy at 25 °C (Table 4). Therefore, aerobic SOM biodecomposition is ruled mainly by the enthalpy of the reactions. Consequently, values for Gibbs energy change in M samples from the UK are less negative than those from Ireland, and the entropy change is lower, suggesting that substrates being catabolized in mineral UK samples are less structurally complex than the rest of samples. These features suggest SOM at different degree of decomposition in all samples.

How these thermodynamic state variables contribute to the variance among samples can be studied by PCA analysis, as shown in Figure 2 and by clustering in Figure 3.

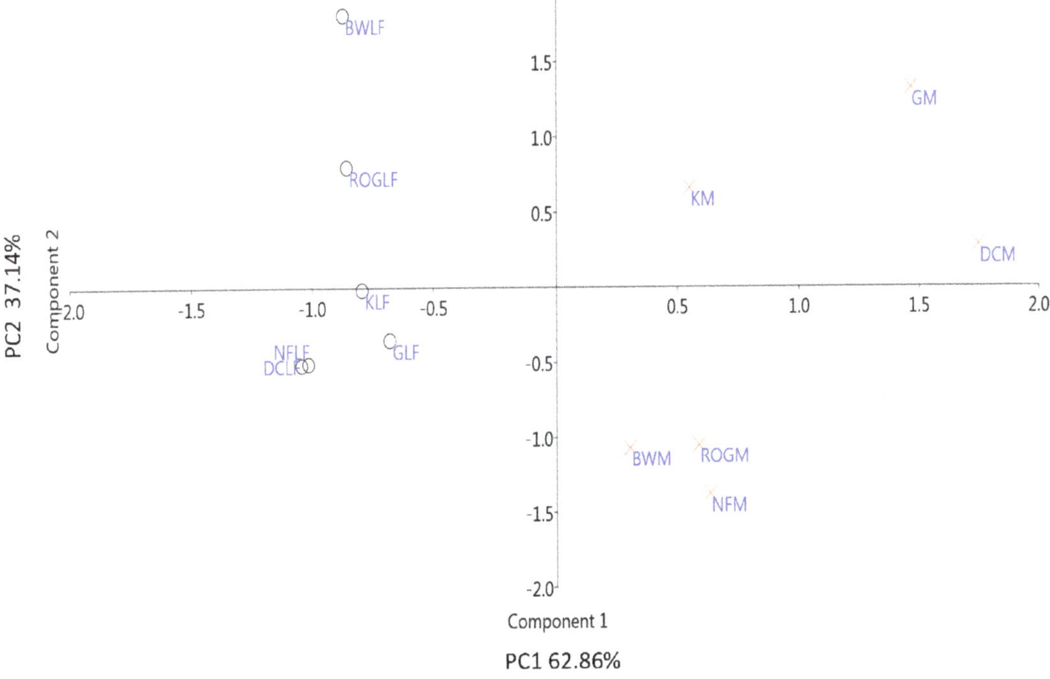


**Figure 2.** PCA plot showing the distribution of soil samples along with the two principal components. Samples ROGLF, BWLF, and NFLF are from the UK. Samples DCLF, GLF, and KLF come from Ireland. All of them represent the soil LF layer and scatter on the left side of the plot. Mineral samples from the UK (ROGM, BWM, and NFM) appear concentrated in the quadrant limited by the positive $x$ and negative $y$-axis. Mineral samples from Ireland (DCM, GM, and KM) scatter along the positive x and y-axis quadrant.

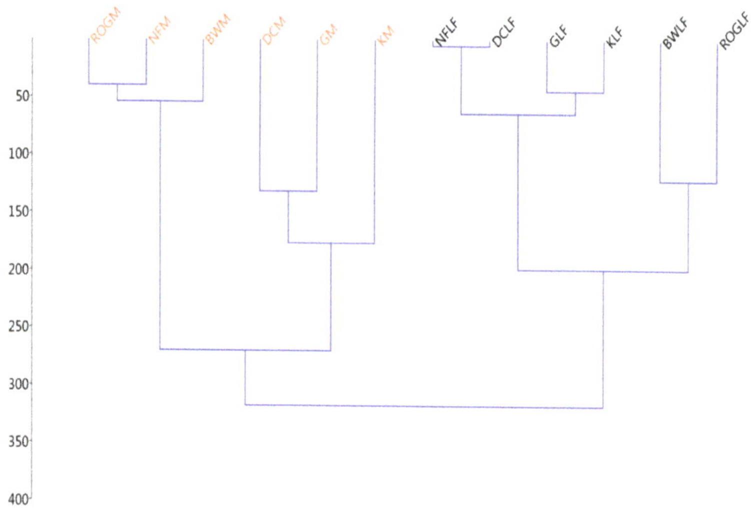


**Figure 3.** Clusters of the soil samples based on their thermodynamic properties. LF samples from the UK (ROGLF, BWLF, and KLF) and Ireland (DCLF, GLF, and KLF) form a cluster separated from two clusters of mineral samples from the UK (ROGM, BWM, and NFM) and Ireland (DCM, GM, and KM).

PCA shows that 99% of the variance can be explained by two principal components, principal component 1 (PC1; 62.86%) and principal component 2 (PC2; 37.14%). The thermodynamic state variables for Reaction (1), representing the SOM thermodynamic properties, are the ones with higher contribution to component 1, while the thermodynamic state variables determined for Reaction (4), representing SOM microbial catabolism, are the ones with higher contribution to component 2. The distribution of samples in the PCA plot in Figure 2 suggests spatial variation in their thermodynamic properties that can be attributed to spatial changes in the composition of soil organic substrates.

Clusters by the thermodynamic state variables can be observed in Figure 3. There is a clear differentiation between the LF layers and the mineral samples. Soil thermodynamic properties vary with soil depth and denote different states of SOM transformation. Mineral samples from the UK and Ireland form two different clusters, while the LF samples are more overlapped.

All these samples come from the same forest ecosystem, oak. Although SOM decomposition originates from oak leaves in all cases, the thermodynamic properties suggest diverse products from the decomposition. This is in agreement with the ecological hypothesis for coexisting microbial species on the same substrate by yielding different end products following thermodynamic constraints. Recent literature reveals how thermodynamics asserts coexistence of various species by the Gibbs energy change available from the metabolic conversions [44]. Results suggest that microbial coexistence, essential to keep microbial diversity, is based on supporting reactions with low Gibbs energy change when using the same substrate. Environments supporting reactions with low Gibbs energy change (more negative) would allow more metabolic diversity than environments allowing reactions at high Gibbs energy change (less negative). When microorganisms start to yield products at similar Gibbs energy as reactants (higher Gibbs energy change), the system would be evolving towards thermodynamic inhibition. Microorganisms overcome this limitation by yielding different products at low concentrations in the environment. Under that theory, mineral soils from the UK would be supporting lower microbial diversity and would be closer to the thermodynamic inhibition (Gibbs energy change closer to zero) than those from Ireland.

The expected natural trend for survival would be by favoring the coexistence of a high number of different metabolic conversions in the environment. The end products of these metabolic conversions are reutilized as an energy source by the existing and/or by the new microbial population, creating a niche for future microbial diversity through adaptation. At the thermodynamic equilibrium ($\Delta G = 0$), two species could not coexist in the same niche, as stated by the competitive exclusion principle [48].

Recent additional work connecting Gibbs energy of dissolved organic matter, DOM, with the microbial diversity [43] also links lower Gibbs energy values to higher microbial diversity.

Microbial biodiversity is given by the number of OTUS of bacteria and fungi (Taxa_S) and by the Shannon index of diversity for four additional mineral soil samples. The mineral soil samples are Leonardite, a recalcitrant material without carbohydrates, a peat sample, and two Cambisols under different management (pine forest and pasture) collected in paired plots in the same location. These samples were surveyed and characterized in a previous paper [47]. Results are reported in Table 5.

**Table 5.** Biological and thermodynamic properties determined for four different mineral soil samples. Microbial diversity was determined by ARISA and is given by the Taxa_S and Shannon_H diversity indices for soil bacteria and fungi. Reproducibility for $-\Delta_c H_{SOM}$ is 5%. The residual standard error is 21.5 kJ mol$^{-1}$. CR values are the average of duplicates with their standard deviation.

| Samples | Taxa_S Bacteria | Taxa_S Fungi | Shannon Bacteria | Shannon Fungi | $-\Delta_c H_{SOM}$ kJmol$^{-1}$C | $-\Delta_c G_{SOM}$ kJmol$^{-1}$C | $-$CR kJmol$^{-1}$$CO_2$-C | $-\Delta_r G_{SOM}$ kJmol$^{-1}$C |
|---|---|---|---|---|---|---|---|---|
| Leonardite | 7 | 21 | 1.92 | 2.07 | 538 | 554 | 273 ± 14 | 276 |
| Pahokee peat | 8 | 15 | 1.93 | 2.19 | 545 | 551 | 463 ± 20 | 468 |
| Cambisol pasture | 28 | 11 | 3.07 | 2.20 | 489 | 495 | 495 ± 21 | 500 |
| Cambisol pine | 6 | 21 | 1.67 | 2.90 | 539 | 544 | 239 ± 18 | 242 |

Leonardite, Pahokee peat, and Cambisol under pine have SOM with $\Delta_c H_{SOM}$ higher than carbohydrates, indicating more reduced substrates. Cambisol under pasture has a lower $\Delta_c H_{SOM}$ (absolute values) than the other samples. Leonardite and Cambisol under pine have less negative $\Delta_r G_{SOM}$ values and their CR values indicate that they are decomposing substrates more oxidized than carbohydrates, representing partial decomposition of substrates or higher component of anaerobic metabolism [31,36]. Their $\Delta_r G_{SOM}$ values indicate microorganisms degrading substrates from SOM at similar free energy to the products. Leonardite contains no carbohydrates in its composition, Cambisol under pine, yes, but shows a CR value similar to the sample without them. A possible explanation could be that carbohydrates are physically protected in the Cambisol under pine, being less available to microorganisms for that reason. Pahokee peat and Cambisol pasture have CR values closer to their $\Delta_c H_{SOM}$ values suggesting degradation of carbohydrates and/or humic material [36]. Peats have a weak mineral matrix and therefore carbohydrates are more available to microorganisms than in Cambisol under pine. That yields lower $\Delta_r G_{SOM}$ (more negative) than the other samples.

Samples with the lowest $\Delta_c G_{SOM}$ are those with SOM more reduced than carbohydrates, with just one exception. Those samples remain the ones with the lowest bacteria diversity (lower Taxa_S values). The Cambisol sample under pasture with $\Delta_c H_{SOM}$ closer to that of carbohydrates is the sample with the highest bacterial diversity, in both TAXA_S and Shannon_H index. The Taxa_S of fungi is lower in the Cambisol sample under pasture than those in the samples with more reduced substrates.

Leonardite and Cambisol pine, with similar CR values and the highest $\Delta_r G_{SOM}$, both have a similar microbiological composition characterized by higher Taxa_S of fungi than that of bacteria. Pahokee peat and Cambisol under pasture have the lowest $\Delta_r G_{SOM}$. Pahokee peat presents a little higher microbial diversity than Leonardite (based on the Shannon_H index) and higher bacteria diversity than Cambisol pine. Although the Cambisol under pasture have the highest bacteria diversity, the low Gibbs energy change of this sample is not reflected in the fungi diversity data.

Figure 4 shows that there are two clear different groups based on the thermodynamic state variables (Figure 4a): the peat and CambisolP1 under pasture also form a different group from those of CambisolPi1 under Pine and Leonardite based on the microbial diversity (Figure 4b).

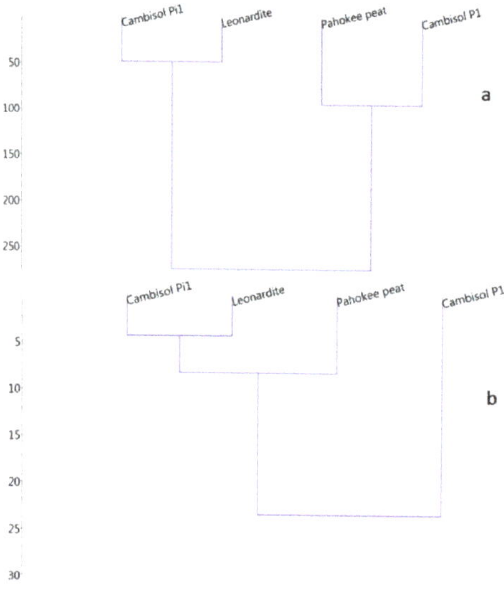


**Figure 4.** The clusters of four mineral soils with differing chemical composition are shown as a function of the thermodynamics state variables (**a**) and as a function of the microbial diversity data in Table 5 (**b**).

Although the set of samples is too low for showing robust results, there is evidence of the possible connection between these thermodynamic variables and the soil microbial composition.

### 3.2.2. Microbial Diversity and Metabolic Carbon Use Efficiency

The measurement of the amount of soil C released to the atmosphere and the amount kept by the soil system represents a way to predict how soil management can contribute to global warming. The scarcity of methodologies to develop these measurements at the microbiological level is responsible for creative alternatives to assess soil microbial metabolic efficiency, defined as the capacity of the soil microbial population to keep C as microbial biomass and that to release it as $CO_2$. Calorimetry is one of the options for that goal. It is considered currently as a method for measuring both metabolic carbon use efficiency, CUE, and metabolic energy use efficiency, EUE, based on previous research showing up applications for soils by calorimetry [39,49] and by calorespirometry [37]. Barros and Feijoo [39] developed a mass and energy balance to assess soil microbial metabolic efficiency for glucose assimilation and compared it with the calorespirometric models [50] to extend the quantification of metabolic CUE to other substrates than glucose. Both models yielded similar values for microbial metabolic efficiency when glucose was used as an external C source [50]. The problem with the calorespirometric models is the limitation to substrates at a lesser degree of reduction than carbohydrates and microbial biomass. This leads to attempts to improve the present thermodynamic models [39] as well as comparisons with other methods [20,40] with satisfactory results for calorimetry. Thermodynamic models for soil CUE and EUE assessment remain a subject that continues evolving at the moment. The existing results provide knowledge about CUE sensitivity to soil management and microbial composition [39,46], the use of inorganic fertilizers [51], or distinct metabolic paths [31]. The role of the soil microbial structure on the metabolic efficiency is explained by studies conducting comparisons between general bacteria and fungi population, demonstrating the importance of the fungi to bacteria ratio on CUE [52], but there is not much information about the metabolic efficiency of numerous substrates taking part in the soil microbial metabolic diversity [42] where calorespirometric procedures could play a valuable role.

## 4. Discussion

The thermodynamic characterization of the soil system will play an interesting role in assessing soil evolution and soil ecology; however, before starting these applications, there is already some tasks with optimization of best procedures to yield the thermodynamic state variables, as well as with the interpretation of them [16,20,26].

The existing methods for giving the redox state of SOM cause debate between the SOM chemical formulation and the direct measurement of the SOM energy content [20,25,53]. Although both of them have pros and cons, they are the best options that we have by now and are a good way to approach the degree of reduction of SOM or the oxidation state (one directly yields the other) to apply them to different soil ecosystems.

The degree of reduction/oxidation of SOM is an option to give a number to the concept of soil recalcitrance (an important goal for soil scientists as one of the procedures to mitigate global warming [18]) and a way to link it to all aspects of soil microbial diversity. It will directly inform us about changes in SOM chemical nature. This paper shows an example of how the degree of reduction indicates the evolution of SOM towards a more reduced state from the soil surface to 5 cm depth and how thermodynamic properties vary among samples from different locations despite sustaining the same forest ecosystem.

The enthalpy of combustion and degree of reduction of SOM give us an idea of the SOM nature through comparisons to well-known organic substrates common in soils such as cellulose, lignin, proteins, and so on. For instance, M samples in Table 2 have SOM more reduced than carbohydrates with values close to those reported for substrates such as lignin, some amino acids, and organic acids [21,53]. This could be explained by the influence on SOM from root exudates at that depth, but also by the SOM natural evolution to a more aromatic state, as reported for original organic matter from oaks [54]. Therefore, an increase in the degree of reduction of SOM with soil depth can be a consequence of a higher degree of SOM transformation or degradation compared to that on the soil surface. SOM in the LF layers has a degree of reduction/oxidation values close to those reported for cellulose, lignocellulosic material, and tree leaves [21,53], compatible with a low degree of SOM decomposition.

Measurement of SOM decomposition by calorespirometry yields the CR of the soil microbial metabolism and the degree of reduction/oxidation of the substrates from SOM being metabolized. By comparing the enthalpies of combustion of SOM to the CR, information about SOM decomposition patterns is obtained. Results show some soils degrade substrates from SOM at a similar degree of reduction as that measured for the entire SOM, while others degrade substrates that can be more, or less, reduced than those of SOM. The possible role of CR to discern metabolic patterns of SOM decomposition is still under development [31]; it could help to support some of the existing theories for SOM decomposition and evolution such as the SOM continuum model [26,55] and to assess other well-known ecological theories such as the maximum power principle [19,21].

The redox state of SOM is the key part of the Gibbs energy change ($\Delta G$) calculation. The role of this thermodynamic variable for soils is still unknown, making it an attractive option to inquire into. For future interpretations, it is essential to consider soil as an open thermodynamic system holding irreversible metabolic microbial reactions. Thus, $\Delta G$ calculations should fix with these premises since the existing equations differentiate between reversible and irreversible processes. Essential, also, is the calorimetric monitoring of the soil microbial reactions because calorimeters are the sole option to detect the exothermic or endothermic nature of those reactions. The heat measured from the soil microbial respiration, the microbial growth reactions, and microbial maintenance are exothermic. The combustion of SOM in the DSC is exothermic, too. Therefore, the enthalpy of combustion and the CR are negative values. $\Delta G$ for any reaction can be positive or negative depending on whether a reaction is, or is not, spontaneous. A negative $\Delta G$ is always expected for spontaneous reactions. The use of the nominative degree of oxidation of SOM or NOSC, to calculate $\Delta G$ involves positive and negative values, too. For this reason, the existing equation to calculate $\Delta G$ from NOSC [16,20] can yield positive $\Delta G$ data. This could be troublesome to interpret. On the contrary, thermodynamic models exerting the degree of reduction of SOM always yield negative $\Delta G$ data. Most of the present models connecting energy and degree of reduction or oxidation include limitations for the metabolism of substrates more reduced than carbohydrates that remain unsolved.

Concerning the role of soil microbial diversity, thermodynamics may play a role in determining the who, where, with whom, and why by the degree of reduction/oxidation of SOM and Gibbs energy. When exploring the microbial structure of the soil samples in this paper, differing degree of reduction of SOM involved changes in microbial diversity. The soil sample at a degree of reduction closer to carbohydrates presented the highest bacterial diversity (Shannon_H index and Tasa_S). Samples with SOM at a higher degree of reduction than carbohydrates showed higher fungi diversity than bacteria, as reported [52]. It seems that bacteria would prefer carbohydrates, while fungi would select SOM at a more substantial degree of decomposition or substrates more reduced than carbohydrates, such as lignin. The varying degree of SOM decomposition would be responsible for the spatial variation of the SOM redox state and the different Gibbs energy values. The reason for this variability is attached to microbial diversity by a type of paper connecting Gibbs energy and microbial diversity to dissolved organic matter, DOM, which solely represents the labile SOM fraction. In this work, Gibbs energy is determined for the entire SOM macromolecule. The preliminary results evidence that Gibbs energy could be sensitive to the soil microbial composition, as reported for DOM. Results in this review suggest, also, that different Gibbs energy values may not accompany greater or lower microbial diversity (structurally and metabolically), but rather changes in the bacterial diversity to fungi diversity relations [52].

This paper also evidences that the same soil type under different management (Cambisol under pine and pasture from paired plots in the same location) yields varied microbial structure and thermodynamic properties. Pasture holds SOM with an enthalpy of combustion close to carbohydrates and lignocellulosic material [31,36], while pine has more reduced SOM than carbohydrates. In this particular case, higher Gibbs energy change of SOM microbial decomposition in pine is attached to higher fungi diversity, while lower Gibbs energy change is obtained for higher bacterial diversity in the pasture. Cambisol under pasture catabolized substrates at a degree of reduction close to that of the entire SOM with metabolic $\Delta G$ values close to those from SOM combustion. Cambisol under pine degrades substrates more oxidized than SOM, yielding higher (less negative) $\Delta G$ values. This could be explained by the different microbial structure, too. Both samples are yielding diverse products from SOM biodecomposition as a consequence of their metabolism and, therefore, building SOM with different thermodynamic properties. This explains the spatial variability of SOM properties [26,44] and supports ecological theories of coexistence based on biodiversity and competitive exclusion principles [48].

## 5. Conclusions

Soil thermodynamic properties appear to be sensitive to soil chemical and biological nature and could be acting as the drivers of the soil properties, defining the shape of the microbial community and their functions. Those functions would evolve by adapting to thermodynamic constraints based on the SOM redox state and the available Gibbs energy. Therefore, to gain accuracy in determining and interpreting the thermodynamic state variables, will be essential to discern strategies for soil survival and soil evolution. The way towards that goal has started.

**Funding:** This research received no external funding.

**Institutional Review Board Statement:** Not applicable.

**Informed Consent Statement:** Not applicable.

**Data Availability Statement:** Thermodynamics of soil organic matter decomposition in semi-natural oak (Quercus) woodland in southwest Ireland—Dryad Digital Repository. https://doi.org/10.5061/dryad.gf1vhhmmd (accessed on 9 July 2020).

**Acknowledgments:** The author thanks Verónica Piñeiro and Montserrat Gómez for DSC–TG and calorespirometric measurements in the RIAIDT_USC analytical facilities of the University of Santiago de Compostela. The author also thanks Elena Vangelova for samples from the Alice Holt Research Station.

**Conflicts of Interest:** The authors declare no conflict of interest.

## References

1. Addiscott, T. Entropy, non-linearity and hierarchy in ecosystems. *Geoderma* **2010**, *160*, 57–63. [CrossRef]
2. Hansen, L.D.; Popović, M.; Tolley, H.D.; Woodfield, B.F. Laws of evolution parallel the laws of thermodynamics. *J. Chem. Thermodyn.* **2018**, *124*, 141–148. [CrossRef]
3. Schrödinger, E. *What Is Life? The Physical Aspects of a Living Cell*; Cambridge University Press: Cambridge, UK, 1944.
4. Odum, E.P. The Strategy of Ecosystem Development. *Science* **1969**, *164*, 262–270. [CrossRef] [PubMed]
5. Fallou, F.A. *First Principles of Soil Science*; G. Schönfield Buchandlung: Dresden, Germany, 1857.
6. Conant, R.T. *Challenges and Opportunities for Carbon Sequestration in Grasslands Systems: A Technical Report on Grassland Management and Climate Change Mitigation*; FAO: Rome, Italy, 2010; Volume 9.
7. Hedges, J.; Eglinton, G.; Hatcher, P.; Kirchman, D.; Arnosti, C.; Derenne, S.; Evershed, R.; Kögel-Knabner, I.; de Leeuw, J.; Littke, R.; et al. The molecularly-uncharacterized component of nonliving organic matter in natural environments. *Org. Geochem.* **2000**, *31*, 945–958. [CrossRef]
8. Tfaily, M.M.; Chu, R.K.; Tolić, N.; Roscioli, K.M.; Anderton, C.; Paša-Tolić, L.; Robinson, E.W.; Hess, N.J. Advanced Solvent Based Methods for Molecular Characterization of Soil Organic Matter by High-Resolution Mass Spectrometry. *Anal. Chem.* **2015**, *87*, 5206–5215. [CrossRef]
9. Šimon, T. Quantitative and qualitative characterization of soil organic matter in the long-term fallow experiment with different fertilization and tillage. *Arch. Agron. Soil Sci.* **2007**, *53*, 241–251. [CrossRef]
10. Ljungholm, K.; Norén, B.; Sköld, R.; Wadsö, I. Use of Microcalorimetry for the Characterization of Microbial Activity in Soil. *Oikos* **1979**, *33*, 15. [CrossRef]
11. Barros, N.; Feijoo, S.; Salgado, J. Calorimetry and soil. *Thermochim. Acta* **2007**, *458*, 11–17. [CrossRef]
12. Suurkuusk, J.; Suurkuusk, M.; Vikegard, P. A multichannel microcalorimetric system: The third generation Thermal Activity Monitor (TAM III). *J. Thermal. Anal. Calorim.* **2017**, *15*, 1–18. [CrossRef]
13. Battley, E.H. *Energetics of Microbial Growth*; John Wiley and Sons: New York, NY, USA, 1987.
14. Heijnen, J.J.; Kleerebezem, R. Bioenergetics of Microbial Growth. In *Encyclopedia of Industrial Biotechnology*; Wiley: Hoboken, NJ, USA, 2010; pp. 1–24.
15. Von Stockar, U.; Maskow, T.; Liu, J.; Marison, I.W.; Patiño, R. Thermodynamics of microbial growth and metabolism: An analysis of the current situation. *J. Biotechnol.* **2006**, *121*, 517–533. [CrossRef] [PubMed]
16. LaRowe, D.E.; Van Cappellen, P. Degradation of natural organic matter: A thermodynamic analysis. *Geochim. Cosmochim. Acta* **2011**, *75*, 2030–2042. [CrossRef]
17. Leifeld, J.; Klein, K.; Wüst-Galley, C. Soil organic matter stoichiometry as indicator for peatland degradation. *Sci. Rep.* **2020**, *10*, 7634–7642. [CrossRef] [PubMed]
18. Schmidt, M.W.I.; Torn, M.; Abiven, S.; Dittmar, T.; Guggenberger, G.; Janssens, I.A.; Kleber, M.; Kögel-Knabner, I.; Lehmann, J.; Manning, D.; et al. Persistence of soil organic matter as an ecosystem property. *Nat. Cell Biol.* **2011**, *478*, 49–56. [CrossRef] [PubMed]
19. Sciuba, E. What did Lotka really say? A critical reassessment of the maximum power principle. *Ecol. Modell.* **2011**, *222*, 1347–1353. [CrossRef]
20. Song, H.-S.; Stegen, J.C.; Graham, E.B.; Lee, J.-Y.; Garayburu-Caruso, V.A.; Nelson, W.C.; Chen, X.; Moulton, J.D.; Scheibe, T.D. Representing Organic Matter Thermodynamics in Biogeochemical Reactions via Substrate-Explicit Modeling. *Front. Microbiol.* **2020**, *11*, 531756. [CrossRef] [PubMed]
21. Gary, C.; Frossard, J.; Chenevard, D. Heat of combustion, degree of reduction and carbon content: 3 interrelated methods of estimating the construction cost of plant tissues. *Agronomie* **1995**, *15*, 59–69. [CrossRef]

22. Jung, H.-J.G.; Varel, V.H.; Weimer, P.J.; Ralph, J. Accuracy of Klason lignin and acid detergent lignin methods as assessed by bomb calorimetry. *J. Agric. Food Chem.* **1999**, *47*, 2005–2008. [CrossRef] [PubMed]
23. Rovira, P.; Henriques, R. Energy content of soil organic matter as studied by bomb calorimetry. *Soil Biol. Biochem.* **2008**, *40*, 172–185. [CrossRef]
24. Baraldi, P.; Beltrami, C.; Cassai, C.; Molinari, L.; Zunarelli, R. Measurements of combustion enthalpy of solids by DSC. *Mater. Chem. Phys.* **1998**, *53*, 252–255. [CrossRef]
25. Malucelli, L.C.; Silvestre, G.F.; Carneiro, J.; Vasconcelos, E.C.; Guiotoku, M.; Maia, C.M.B.F.; Filho, M.A.S.C. Biochar higher heating value estimative using thermogravimetric analysis. *J. Therm. Anal. Calorim.* **2019**, *139*, 2215–2220. [CrossRef]
26. Barros, N.; Fernandez, I.; Byrne, K.A.; Jovani-Sancho, A.J.; Ros-Mangriñan, E.; Hansen, L.D. Thermodynamics of soil organic matter decomposition in semi-natural oak (*Quercus*) woodland in southwest Ireland. *Oikos* **2020**, *129*, 1632–1644. [CrossRef]
27. Roels, J.A. *Energetics and Kinetics in Biotechnology*; Elsevier: Amsterdam, The Netherlands, 1983.
28. Sandler, S.I.; Orbey, H. On the thermodynamics of microbial growth processes. *Biotechnol. Bioeng.* **1991**, *38*, 697–718. [CrossRef]
29. Ludovisi, A.; Pandolfi, P.; Taticchi, M.I. The strategy of ecosystem development: Specific dissipation as an indicator of ecosystem maturity. *J. Theor. Biol.* **2005**, *235*, 33–43. [CrossRef]
30. Battley, E.H. An empirical method for estimating the entropy of formation and the absolute entropy of dried microbial biomass for use in studies on the thermodynamics of microbial growth. *Thermochim. Acta* **1999**, *326*, 7–15. [CrossRef]
31. Chakrawal, A.; Herrmann, A.M.; Šantrůčková, H.; Manzoni, S. Quantifying microbial metabolism in soils using calorespirometry—A bioenergetics perspective. *Soil Biol. Biochem.* **2020**, *148*, 107945. [CrossRef]
32. Maskow, T.; Kemp, R.B.; Buchholz, F.; Schubert, T.; Kiesel, B.; Harms, H. What heat is telling us about microbial conversions in nature and technology: From chip- to megacalorimetry. *Microb. Biotechnol.* **2009**, *3*, 269–284. [CrossRef] [PubMed]
33. Wadsö, L.; Hansen, L.D. Calorespirometry of terrestrial organisms and ecosystems. *Methods* **2015**, *76*, 11–19. [CrossRef]
34. Barros, N.; Feijóo, S.; Hansen, L.D. Calorimetric determination of metabolic heat, $CO_2$ rates and the calorespirometric ratio of soil basal metabolism. *Geoderma* **2011**, *160*, 542–547. [CrossRef]
35. Herrmann, A.M.; Bölscher, T. Simultaneous screening of microbial energetics and $CO_2$ respiration in soil samples from different ecosystems. *Soil Biol. Biochem.* **2015**, *83*, 88–92. [CrossRef]
36. Barros, N.; Hansen, L.; Piñeiro, V.; Pérez-Cruzado, C.; Villanueva, M.; Proupín, J.; Añón, J.A.R. Factors influencing the calorespirometric ratios of soil microbial metabolism. *Soil Biol. Biochem.* **2016**, *92*, 221–229. [CrossRef]
37. Hansen, L.D.; Macfarlane, C.; McKinnon, N.; Smith, B.N.; Criddle, R.S. Use of calorespirometric ratios, heat per $CO_2$ and heat per $O_2$, to quantify metabolic paths and energetics of growing cells. *Thermochim. Acta* **2004**, *422*, 55–61. [CrossRef]
38. Barros, N. Calorimetry and soil biodegradation: Experimental procedures and thermodynamic models. In *Toxicity and Biodegradation Testing*; Humana Press: Totowa, NJ, USA, 2018; pp. 123–145.
39. Harris, J.A.; Ritz, K.; Coucheney, E.; Grice, S.; Lerch, T.Z.; Pawlett, M.; Herrmann, A.M. The thermodynamic efficiency of soil microbial communities subject to long-term stress is lower than those under conventional input regimes. *Soil Biol. Biochem.* **2012**, *47*, 149–157. [CrossRef]
40. Geyer, K.M.; Dijkstra, P.; Sinsabaugh, R.; Frey, S.D. Clarifying the interpretation of carbon use efficiency in soil through methods comparison. *Soil Biol. Biochem.* **2019**, *128*, 79–88. [CrossRef]
41. Chakrawal, A.; Herrmann, A.M.; Manzoni, S. Leveraging energy flows to quantify microbial traits in soils. *Soil Biol. Biochem.* **2021**, *155*, 108169. [CrossRef]
42. Xu, J.; Feng, Y.; Barros, N.; Zhong, L.; Chen, R.; Lin, X. Exploring the potential of microcalorimetry to study soil microbial metabolic diversity. *J. Therm. Anal. Calorim.* **2016**, *127*, 1457–1465. [CrossRef]
43. Zhang, J.; Feng, Y.; Wu, M.; Chen, R.; Li, Z.; Lin, X.; Zhu, Y.; Delgado-Baquerizo, M. Evaluation of Microbe-Drive Soil Organic Matter Quantity and Quality by Thermodynamic Theory. *mBIO* **2021**, *12*, e03252-20. [CrossRef]
44. Grosskopf, T.; Soyer, O.S. Microbial diversity arising from thermodynamic constrains. *ISME J.* **2016**, *10*, 2725–2733. [CrossRef]
45. Hansen, L.D.; Tolley, H.D.; Woodfield, B.F. Tranformation of matter in living organisms during growth and evolution. *Biophys. Chem.* **2021**, 106550. [CrossRef] [PubMed]
46. Bölscher, T.; Paterson, E.; Freitag, T.; Thornton, B.; Herrmann, A.M. Temperature sensitivity of substrate-use efficiency can result from altered microbial physiology without change to community composition. *Soil Biol. Biochem.* **2017**, *109*, 59–69. [CrossRef]
47. Hansen, L.D.; Barros, N.; Transtrum, M.K.; Rodríguez-Añón, J.A.; Proupín, J.; Piñeiro, V.; Arias-González, A.; Gartzia-Bengoetxea, N. Effect of extreme temperatures on soil: A calorimetric approach. *Thermochim. Acta* **2018**, *670*, 128–135. [CrossRef]
48. Hardin, G. The Competitive Exclusion Principle. *Science* **1960**, *131*, 1292–1297. [CrossRef]
49. Barros, N.; Feijóo, S. A combined mass and energy balance to provide bioindicators of soil microbiological quality. *Biophys. Chem.* **2003**, *104*, 561–572. [CrossRef]
50. Barros, N.; Salgado, J.; Rodríguez-Añón, J.A.; Proupín, J.; Villanueva, M.; Hansen, L.D. Calorimetric approach to metabolic carbón conversión efficiency in soils: Comparison of experimental and theoretical models. *J. Thermal. Anal. Calorim.* **2010**, *99*, 771–777. [CrossRef]
51. Barros, N.; Feijoo, S.; Simoni, J.A.; Airoldi, C.; Ramajo, B.; Espina, A.; García, J.R. A mass and energy balance to provide microbial growth yield efficiency in soil: Sensitivity to metal layering phospates. *J. Therm. Anal. Calorim.* **2008**, *93*, 657–665. [CrossRef]
52. Soares, M.; Rousk, J. Microbial growth and carbon use efficiency in soil: Links to fungal-bacterial dominance, SOC-quality and stoichiometry. *Soil Biol. Biochem.* **2019**, *131*, 195–205. [CrossRef]

53. Masiello, C.A.; Gallagher, M.E.; Randerson, J.T.; Deco, R.M.; Chadwick, O.A. Evaluating two experimental approaches for measuring ecosystem carbon oxidation state and oxidative ratio. *J. Geophys. Res. Space Phys.* **2008**, *113*, 03010. [CrossRef]
54. Chavez-Vergara, B.; Merino, A.; Vázquez-Marrufo, G.; García-Oliva, F. Organic matter dynamics and microbial activity during decomposition of forest floor under two native neotropical oak species in a temperate deciduous forest in Mexico. *Geoderma* **2014**, *235–236*, 133–145. [CrossRef]
55. Lehmann, J.; Kleber, M. The contentious nature of soil organic matter. *Nat. Cell Biol.* **2015**, *528*, 60–68. [CrossRef]

MDPI
St. Alban-Anlage 66
4052 Basel
Switzerland
Tel. +41 61 683 77 34
Fax +41 61 302 89 18
www.mdpi.com

*Applied Sciences* Editorial Office
E-mail: applsci@mdpi.com
www.mdpi.com/journal/applsci

www.ingramcontent.com/pod-product-compliance
Lightning Source LLC
LaVergne TN
LVHW070615170726
843515LV00004B/84

* 9 7 8 3 0 3 6 5 5 1 4 9 4 *